The EU AI Act and Global AI Compliance

A Step-By-Step Regulatory Playbook

Dr. C Louis-Cha

Cybersoft Publishing LLC

Fort Washington, MD 20744

Drclaude.net

First Edition April 2026

1 Table of Contents

2 Introduction

This book is organized around a single practical question: what does it take to lead an enterprise through AI compliance in an era of enforceable, evolving, globally fragmented regulation? It is not a legal reference. It is not an academic survey of AI ethics frameworks. It is a manager's guide to the decisions, workflows, and organizational structures that determine whether an enterprise meets its compliance obligations and does so in a way that builds durable capability rather than generating paperwork.

The answer to that central question unfolds across nine chapters, each addressing a distinct dimension of the challenge: from understanding the regulatory architecture itself to building the internal function capable of sustaining compliance over time. The chapters are designed to be read in sequence, because each one builds on what came before. But they are also written to stand alone as references for specific decisions, so that a leader facing a conformity assessment question in Chapter Five does not need to re-read Chapters One through Four to find actionable guidance.

What makes this book different from the legal and consulting literature on the EU AI Act is its orientation. The law itself is written for regulators and courts. Legal commentary is written for lawyers. This book is written for the manager who has to make the law real inside an organization: who has to explain it to a board, negotiate it with a product team, budget for it with a CFO, and defend it to an auditor. The orientation throughout is practical, the language is plain, and every concept is anchored to the decisions and workflows it actually affects.

A note on scope: the EU AI Act is the spine of this book, but it is not the whole skeleton. Chapter Three addresses the global regulatory landscape in depth, covering US state laws, Canada's AIDA framework, Brazil, and China. Subsequent chapters weave in cross-jurisdictional considerations wherever they affect managers' compliance decisions. The goal is operational clarity across the full compliance environment, not an artificially narrow focus on a single jurisdiction.

How This Book Is Organized

The nine chapters of this book are organized into three clusters, each addressing a different phase of the compliance journey. The first cluster (Chapters One and Two) establishes context. The second cluster (Chapters Three through Seven) widens the lens to global compliance and then narrows it to the operational specifics that matter most. The third cluster (Chapters Eight and Nine) addresses organizational design and strategic durability, the capabilities that allow compliance to endure as the regulatory environment continues to evolve.

Diagram I.1 - Book Roadmap

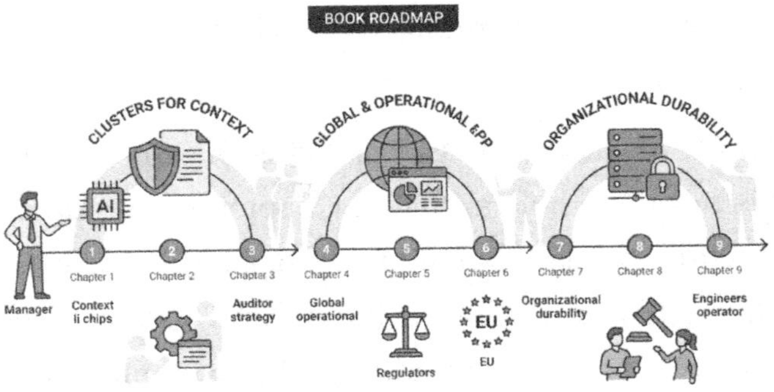

A strategic framework for digital resilience and innovation.

The journey begins with context. Chapter One situates the EU AI Act within the broader shift from voluntary AI principles to binding legal obligation, explaining why regulators moved when they did and what the enforcement architecture looks like. Chapter Two maps the Act's internal structure: its risk tiers, its prohibited practices, its obligations for high-risk systems, and its emerging rules for general-purpose AI, so that readers can locate their own products and operations within the framework before engaging with the compliance specifics that follow.

Chapters Three through Five widen the lens, then return to operational specifics. Chapter Three surveys the global regulatory landscape beyond Europe, covering US state laws, Canada, Brazil, China, and the challenge of maintaining coherent compliance across multiple overlapping jurisdictions. Chapter Four confronts what compliance actually requires in practice: the people, processes, and technology investments that transform regulatory requirements into organizational capability. Chapter Five addresses the conformity assessment process, the documentation obligations, and the role of technical standards and third-party auditors in establishing and demonstrating compliance. Readers managing the pilot-to-production pathway for a high-risk AI system will find Chapter Five particularly concrete.

Chapters Six and Seven go deep on the two categories of AI that carry the heaviest regulatory burden. Chapter Six examines high-risk AI systems, covering their classification, their specific obligations, quality management requirements, and the demands of post-market monitoring. Chapter Seven turns to general-purpose AI and foundation models, addressing transparency obligations, systemic risk rules, and the question of downstream liability when a foundation model is integrated into another organization's product. The final two chapters address organizational design and strategic durability. Chapter Eight covers building the compliance function itself: roles, reporting structures, cross-functional coordination mechanisms, and budget realities. Chapter Nine closes the analytical journey by asking how an organization maintains compliance as the law continues to evolve, treating regulatory interoperability, horizon scanning, and adaptive governance not as afterthoughts but as core capabilities for any enterprise that intends to operate in the AI economy for the long term.

Key Terms at a Glance

Before the chapters begin, it is worth establishing the core vocabulary that runs throughout the book. These are not legal definitions; they are working definitions that give a manager enough precision to follow the compliance logic without requiring a law degree. The EU AI Act organizes its obligations around a four-tier risk classification. At the top of the hierarchy are prohibited practices: AI applications that the Act bans outright because their potential for harm is judged to be unacceptable under any circumstances, including subliminal manipulation, social scoring by public authorities, and most forms of real-time biometric surveillance in public spaces. Below that tier sit high-risk systems, which face the most detailed compliance obligations. High-risk designation applies to AI used in hiring and recruitment, credit and insurance decisions, educational assessment, critical infrastructure management, law enforcement, judicial proceedings, and border control, among other domains. Limited-risk systems face targeted transparency requirements (for example, an obligation to disclose when a user is interacting with a chatbot) but not the full compliance burden of high-risk classification. Minimal-risk systems, which include most consumer AI applications, face no specific obligations under the Act.

Conformity assessment is the structured process by which an organization demonstrates that its AI system meets the Act's requirements before placing it on the market or putting it into service. For most high-risk systems, this can be conducted through the Act's self-assessment pathway, which imposes demanding documentation and testing requirements. For systems in certain categories (critical infrastructure, law enforcement, migration management), third-party assessment bodies must be involved. Successful completion of conformity assessment allows a provider to affix the CE marking to its product, indicating conformity with EU law and enabling market access across EU member states. Under the Act's framework, providers are the organizations that develop or place an AI system on the market. Deployers are the organizations that put a provider's system into use in their own operations. The distinction matters because providers and deployers carry different obligations, and the allocation of responsibility between them is one of the most practically important questions a manager faces when an AI system crosses organizational boundaries.

Post-market monitoring refers to the ongoing obligation to track the performance of a deployed high-risk AI system, detect problems, and report serious incidents to supervisory authorities within defined timeframes. General-purpose AI (GPAI) refers to AI models trained on broad data that can perform a wide range of tasks and may be integrated into many different downstream applications. The Act imposes baseline transparency obligations on all GPAI providers and heightened requirements on those whose models are designated as posing systemic risk. Finally, the fundamental rights impact assessment is a structured evaluation that deployers of certain high-risk AI systems are required to conduct before deployment, examining how the system may affect the rights of individuals it interacts with or makes decisions about. These terms form the vocabulary of responsible AI governance, and the chapters that follow give each of them operational meaning.

Diagram I.2 - Core Compliance Vocabulary

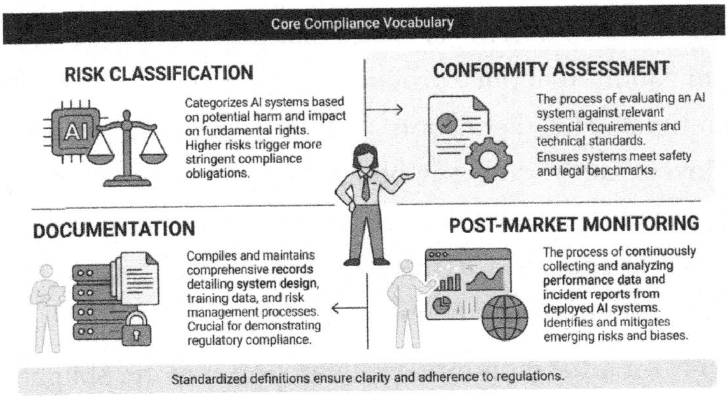

Core Compliance Vocabulary
RISK CLASSIFICATION
Categorizes AI systems based on potential harm and impact on fundamental rights. Higher risks trigger more stringent compliance obligations.
AI
CONFORMITY ASSESSMENT
The process of evaluating an AI system against relevant essential requirements and technical standards. Ensures systems meet safety and legal benchmarks.
DOCUMENTATION
Compiles and maintains comprehensive records detailing system design, training data, and risk management processes. Crucial for demonstrating regulatory compliance.
POST-MARKET MONITORING
The process of continuously collecting and analyzing performance data and incident reports from deployed AI systems. Identifies and mitigates emerging risks and biases.
Standardized definitions ensure clarity and adherence to regulations.

3 When Principles Became Law

A senior vice president of technology at a European retail bank spent most of 2022 confident that her organization's AI governance program was ahead of the curve. The bank had adopted a published set of AI principles, established an ethics review board, and required that internal model development teams complete a responsible-AI checklist before any system went into production. When the EU AI Act's risk classification framework was explained to her legal team in early 2023, the initial reaction was reassurance: surely a governance program this deliberate would map neatly onto whatever the regulators required. By late 2023, that confidence had given way to something more uncomfortable. The bank's credit-scoring models, its fraud-detection systems, and its customer-service automation tools all appeared to qualify as high-risk AI systems under the Act's Annex classifications. That meant documented technical files, mandatory conformity assessments, human oversight mechanisms, and a post-market monitoring program - none of which the existing ethics checklist had been designed to produce. The gap between principled intention and legally enforceable obligation proved very wide.

The bank's experience is not unusual. Organizations across financial services, healthcare, recruitment, and public-sector contracting are discovering that voluntary governance frameworks, however thoughtfully constructed, were built for a different environment. The EU AI Act created an environment where the question is no longer 'Does our AI align with our values?' but rather 'Can we demonstrate to a supervisory authority that this system meets every applicable obligation?' That is a different question, and it requires a different kind of answer. This chapter explains how the regulatory environment shifted, why it shifted when it did, and what the shift means for the decisions enterprise leaders must now make.

3.1 Why It Matters

For a manager responsible for AI deployment, the shift from voluntary ethics frameworks to enforceable law changes the nature of every decision in the pipeline. Under the voluntary era, a team that deployed an AI system with incomplete bias testing faced reputational risk - the possibility of public criticism, internal review, or withdrawal of the system if problems emerged. Under binding law, the same team faces regulatory exposure: fines, market access restrictions, mandatory remediation orders, and, in some jurisdictions, personal liability for senior executives who signed off on non-compliant deployments. The risk category has changed, and risk categories determine where governance resources get allocated.

The practical consequence is that AI compliance cannot be delegated entirely to an ethics committee or a data science team. Boards and executive committees are now in the accountability chain. Audit functions are being asked to treat AI compliance as a standing audit domain. Legal and procurement teams are writing AI-specific clauses into vendor contracts. These changes do not happen because organizations suddenly became more ethical. They happen because the law created accountability structures that did not exist before, and accountability structures change behavior in ways that voluntary commitments rarely do.

Understanding the historical arc that produced binding AI law is mission-aligned work for any manager building a compliance program. Regulators did not arrive at their current position by accident. They followed a logic shaped by specific failures, specific political dynamics, and specific lessons drawn from prior regulatory projects. A manager who understands that logic is better positioned to anticipate how the framework will evolve, where supervisory authorities will focus their enforcement attention first, and what kinds of governance investments will prove durable over a multi-year compliance horizon.

3.2 The Limits of the Voluntary Era

Between roughly 2016 and 2021, the dominant mode of AI governance was the publication of principles. Technology companies, industry associations, national governments, and international bodies have produced an extensive body of literature on AI ethics frameworks, each identifying values such as fairness, transparency, accountability, and human oversight as goals for AI development. The exercise had genuine intellectual merit. It established a shared vocabulary, surfaced important questions about how algorithmic systems affect human lives, and created internal pressure within some organizations to think more carefully about system design. What it did not produce were consistent, verifiable, enforceable outcomes.

Why Principles-Based Governance Failed to Produce Consistent Outcomes

The fundamental problem with principles-based AI governance was that it asked organizations to self-define what compliance looked like. A company that had adopted a 'fairness' principle could satisfy that commitment with a two-paragraph policy statement, a statistical parity check on a single output variable, or a full algorithmic impact assessment with an external audit. All three could plausibly be described as forms of fairness governance, and no external body had the authority to determine which was adequate. The result was a governance landscape characterized by wide variation in rigor, with market incentives generally rewarding faster deployment over more thorough scrutiny.

Research produced during this period consistently found that AI ethics principles, when examined across organizations, shared rhetorical similarity while diverging sharply in implementation. The same word - 'transparency' - covered systems that disclosed a model's existence to users, systems that published high-level descriptions of training data, and systems that provided full technical documentation, including architecture, hyperparameters, and performance metrics across demographic subgroups. A manager trying to benchmark her organization against peers found that the benchmarks were essentially self-reported and self-defined. This made it almost impossible to identify which organizations were genuinely managing AI risk and which were engaged in what observers began calling 'ethics washing.'

The gap also showed up in organizational structure. Ethics review boards, where they existed, typically operated as advisory bodies without formal authority to block deployments. Model risk functions in financial services were often technically sophisticated but narrowly focused on quantitative model performance rather than the broader social consequences of system outputs. Legal teams were generally not engaged until a specific legal risk had been identified, which meant they were rarely involved in the upstream design decisions where consequential choices were made. The result was that even in organizations with genuine ethical commitment, governance structures were not positioned to catch the problems that AI ethics principles were designed to prevent.

Diagram 1.1 - The Voluntary Governance Gap

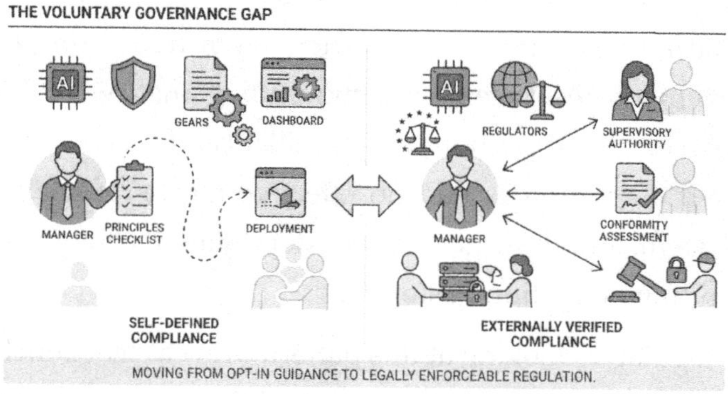

The Accountability Gap Between AI Developers and Affected Communities

Principles-based governance also failed to resolve a structural accountability problem: the people designing and deploying AI systems and the people affected by those systems existed in almost entirely separate accountability relationships. A person denied a loan by an automated credit-scoring system had no reliable mechanism to learn that AI was involved in the decision, no right to a human review of that decision, and no avenue to challenge the system's design or training data. The organization deploying the system had accepted some reputational risk by publishing an ethics statement, but faced no binding obligation to provide explanations, enable appeals, or disclose how the system was built.

This accountability gap was not merely a procedural concern. It produced outcomes that were difficult to defend on any ethical standard. Facial recognition systems deployed by law enforcement agencies in several jurisdictions produced significantly higher error rates for women and people with darker skin tones. Automated hiring tools trained on historical promotion data reproduced historical biases against women in technical roles. Predictive policing algorithms concentrated enforcement resources in communities that were already subject to disproportionate surveillance. In each case, the affected communities had no effective means of challenging the system, and the organizations deploying the systems faced no binding obligation to stop.

Regulators looking at this pattern drew a clear conclusion: voluntary commitments had not closed the accountability gap because closing the gap required giving affected individuals rights that could be exercised against deploying organizations, and rights require law, not principles. The EU AI Act's provisions for human oversight of high-risk systems, the right-to-explanation provisions inherited and extended from the GDPR, and the requirement that providers and deployers maintain technical documentation accessible to supervisory authorities are all direct responses to the accountability gap that the voluntary era failed to close.

Diagram 1.2 - Accountability Gap Under Voluntary Governance

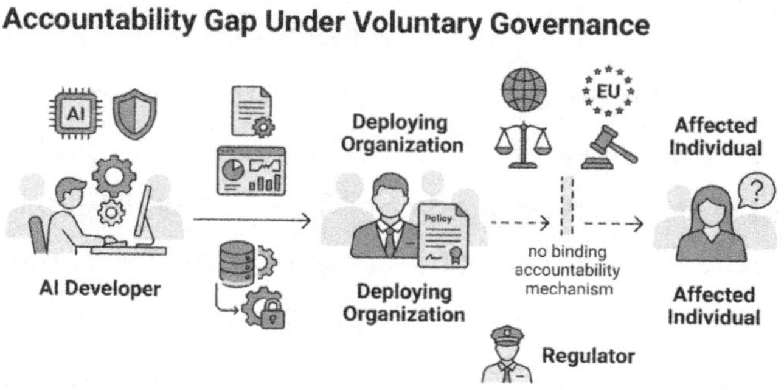

Accountability Gap undernar Voluntary Governance — no binding accountability mechanism le closers.

High-Profile Failures That Accelerated the Case for Binding Rules

The political case for binding AI law was built, in part, on a set of documented failures that received significant public and parliamentary attention in the years leading up to the passage of the EU AI Act. These failures shared a common structure: an AI system deployed in a consequential domain produced outcomes that harmed identifiable individuals or communities, and the organization responsible for the system lacked either the information or the incentive to correct the problem before it caused substantial harm. The failures were not hypothetical or speculative. They involved real credit decisions, real hiring outcomes, real criminal risk scores, and real facial recognition identifications that sent people to prison based on erroneous matches.

Each high-profile failure added pressure to the political case that self-regulation was insufficient. The argument that 'the technology industry can be trusted to govern itself' became harder to sustain when the same period that produced ethics principles also produced documented cases of biased hiring tools being sold to major corporations, fraudulent claims about system capabilities by AI vendors, and organizations that quietly discontinued AI systems after internal reviews revealed problems without disclosing those problems to affected individuals or regulators. The credibility gap between stated principles and actual practice became one of the central arguments regulators used to justify the move to binding law.

For enterprise managers, these cases carry a lesson that goes beyond their historical interest. The kinds of failures that accelerated binding AI law - systematic bias in consequential decisions, inadequate disclosure to affected individuals, governance structures that caught problems after deployment rather than before - are not artifacts of a less sophisticated era. They are the natural outcome of deploying complex systems under resource constraints and competitive pressure, without adequate compliance infrastructure in place. The EU AI Act's requirements for data governance, human oversight, and post-market monitoring are designed to address the conditions that led to those failures. A manager who understands that design logic is better equipped to build a compliance program that addresses the actual risk, not just the technical checklist.

Diagram 1.3 - Failure Patterns That Drove Regulation

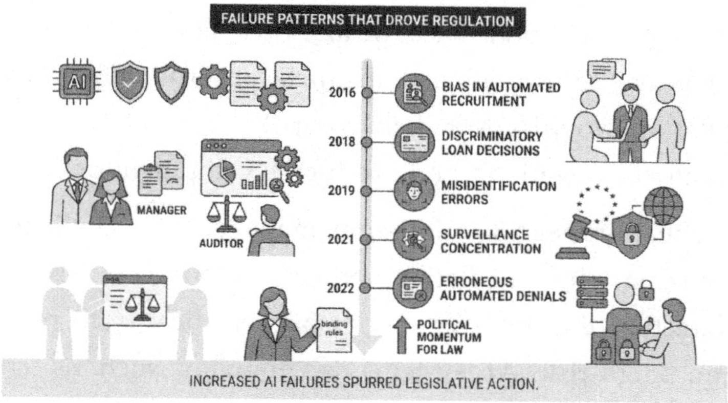

3.3 The Political Economy of AI Regulation

Binding AI law did not emerge from a single policy argument. It emerged from a convergence of institutional interests, documented evidence of harm, GDPR's demonstrated enforceability, and a political environment in which European leaders saw digital regulation as an expression of values-based governance. Understanding the political economy that produced the EU AI Act helps managers understand both what the law is trying to accomplish and where its enforcement priorities are likely to cluster.

How the European Union Built Consensus for a Horizontal AI Law

The EU AI Act is what regulators call a horizontal law: it applies across sectors rather than targeting a single industry. This is a significant design choice. Horizontal legislation is more complex to draft than sector-specific rules because it must accommodate the full range of AI applications, from medical diagnosis to credit scoring to content recommendation. It also requires building political consensus across industry groups with very different interests and risk profiles. The pharmaceutical, financial services, and automotive sectors all had distinct concerns about how AI regulation would affect their competitive positions, existing regulatory obligations, and product development timelines.

The European Commission managed this challenge by organizing the legislative process around a risk-based architecture that tied regulatory burden directly to potential harm. High-risk AI systems in domains such as critical infrastructure, employment, education, and law enforcement would face the most stringent requirements. Lower-risk applications would face lighter obligations or none at all. This framework gave industry stakeholders a way to engage constructively with the legislative process: they could argue about where the risk boundaries should sit without having to argue against the principle of proportionate regulation. The result was a law that survived years of negotiation without losing its structural logic.

The process also benefited from the institutional experience European regulators had gained through the implementation of the GDPR. By the time AI Act negotiations reached their final stages, data protection authorities across Europe had developed supervisory capacity, enforcement precedents, and interpretive guidance that demonstrated the EU's ability to operationalize complex digital regulation. That track record reduced the credibility of industry arguments that the EU AI Act was unworkable in practice. It also gave European policymakers confidence that the supervisory infrastructure needed to enforce the Act could be built, even if it took time to mature.

Diagram 1.4 - EU AI Act Legislative Path

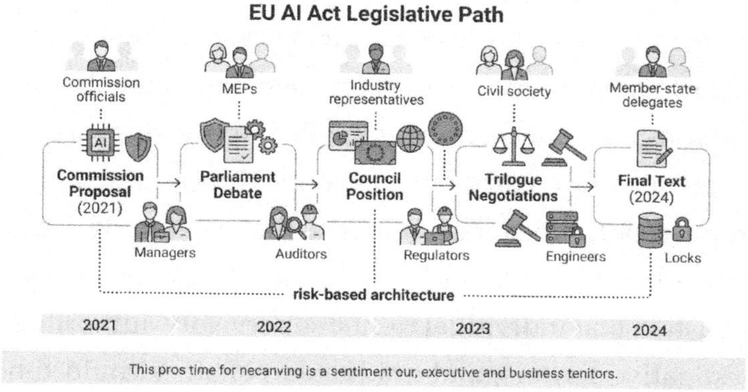

The Role of the GDPR Precedent in Shaping Regulatory Ambition

No analysis of the EU AI Act's political context is complete without understanding how the General Data Protection Regulation changed what European regulators believed was possible. GDPR, which entered into force in May 2018, was itself a significant leap: it applied extraterritorially to organizations processing data about EU residents regardless of where those organizations were established, it imposed substantial fines (up to four percent of global annual turnover for major violations), and it created a comprehensive set of individual rights including access, rectification, erasure, and portability. Before GDPR took effect, the conventional wisdom in the technology industry was that European regulators would struggle to enforce such an ambitious regulation against large non-EU platforms. That conventional wisdom proved wrong.

Several major enforcement actions against non-EU technology companies demonstrated that the GDPR's extraterritorial reach was operational rather than merely theoretical. Data protection authorities developed the capacity to investigate complex cases involving cross-border data flows, to coordinate enforcement across member states through the One-Stop-Shop mechanism, and to impose fines significant enough to affect corporate behavior. These outcomes gave European policymakers confidence that a similarly ambitious AI regulation could achieve similar results. They also established that individual rights, supervisory authorities with investigative powers, and graduated penalties could function as effective regulatory tools in the digital domain.

For enterprise managers, the GDPR precedent also serves as a calibration tool for thinking about AI Act enforcement timelines. GDPR's first major enforcement actions came two to three years after the regulation entered into force. The early enforcement phase was characterized by cases involving clear, demonstrable violations - unlawful data processing, inadequate consent mechanisms, and data breach notifications that were late or incomplete. Managers who understand this trajectory can anticipate that EU AI Act enforcement is likely to begin with the most clearly documented non-compliance: organizations that deploy systems in the high-risk categories without producing the required technical documentation, without implementing human oversight mechanisms, or without registering in the EU database for high-risk systems. Building compliance with those foundational requirements first is both the legally correct priority and the strategically sensible one.

Diagram 1.5 - GDPR to AI Act Regulatory Logic

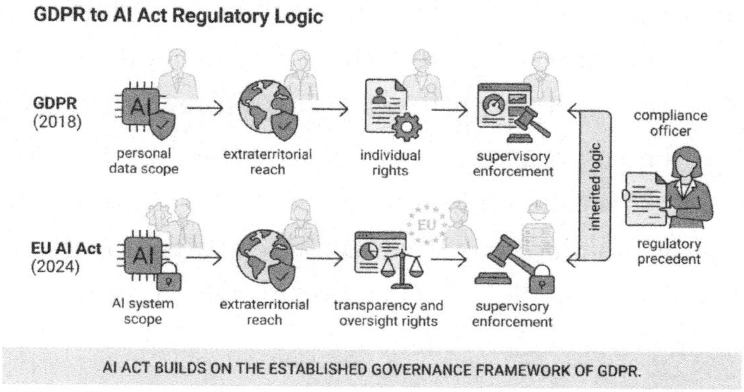

Industry Responses and the Limits of Self-Regulatory Proposals

During the legislative process, industry coalitions put forward several self-regulatory alternatives, arguing that binding law was premature given the pace of technological change and the difficulty of writing durable rules for a rapidly evolving domain. These proposals generally shared a common structure: a voluntary commitment to adopt standards of internal governance, an industry-led certification scheme, and a review mechanism to update standards as technology evolved. The proposals were not cynical exercises in delay, at least not uniformly. Some came from organizations that had invested substantially in responsible-AI practices and were genuinely concerned that prescriptive legislation would lock in standards that would quickly become obsolete.

Regulators rejected the self-regulatory path for several reasons that are worth understanding. First, the voluntary era had already demonstrated that self-regulatory commitments produced inconsistent outcomes. Second, the competitive dynamics of AI markets created strong pressures to minimize compliance costs, which meant that even organizations with genuine commitments would face difficulty sustaining them under competitive pressure without binding requirements that applied equally to competitors. Third, self-regulatory schemes lacked the enforcement mechanisms needed to give affected individuals meaningful rights. A complaint about a vendor's voluntary standard could be submitted to the vendor's own compliance office. Still, it had no pathway to a court, a fine, or a binding remediation order.

The residue of the self-regulatory debate is evident in the structure of the EU AI Act in several places. The Act includes extensive provisions for harmonized standards, which allow industry bodies to develop detailed technical specifications that, once adopted, create a presumption of compliance. This channel gives industry genuine input into the practical content of the regulatory framework while maintaining the law's binding authority. It is a hybrid model that reflects the compromise between regulatory ambition and industry concern about prescriptive rigidity. For managers, harmonized standards are a priority monitoring item: adopting them early, when they become available, is likely the most practical way to demonstrate conformity for most high-risk system deployments.

Diagram 1.6 - Self-Regulation vs. Binding Law Tradeoffs

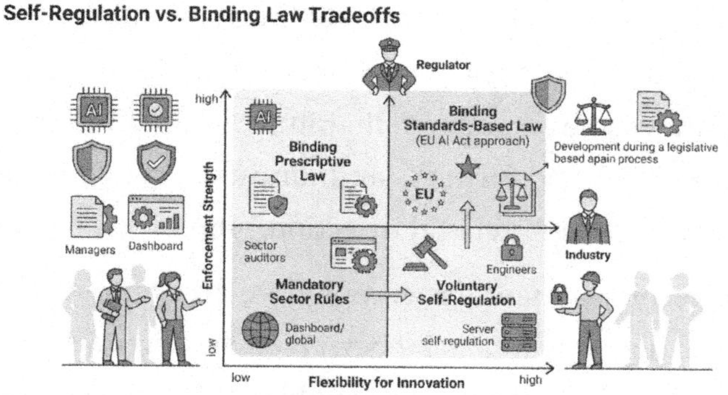

Optimizing compliance and growth in the digital economy.

3.4 Jurisdictional Spread and the Race for Regulatory Leadership

The EU AI Act did not emerge in a regulatory vacuum. By the time the Act was finalized, multiple jurisdictions were either enacting or actively developing binding AI frameworks. This near-simultaneous regulatory movement reflects a common set of inputs - the same documented failures, similar political pressures, and growing recognition that voluntary approaches were insufficient - filtered through very different institutional traditions and policy priorities. The result is a global regulatory landscape that enterprise leaders must navigate as a whole rather than as a series of isolated local requirements.

Why Multiple Jurisdictions Moved Toward Binding AI Rules Simultaneously

The convergence of AI regulatory activity across jurisdictions in the 2022-2024 period reflects several reinforcing dynamics. The first is the availability of shared evidence: the same research on algorithmic bias, the same documented failures of facial recognition systems, and the same analyses of risks from large language models were circulating among policymakers in Washington, Ottawa, London, Brasilia, Beijing, and Singapore simultaneously. Regulators were reading the same academic literature, attending the same international policy forums, and drawing conclusions that were, at least in their general direction, similar.

The second dynamic is competitive and reputational. No major jurisdiction wants to be perceived as a permissive haven for harmful AI while its neighbors build governance frameworks. The EU's move toward binding law created political pressure on other jurisdictions to demonstrate comparable regulatory seriousness, even when their specific approaches differed substantially. Canada's Artificial Intelligence and Data Act, the United Kingdom's sector-specific pillar approach, and several US state laws all emerged in a political environment where 'we are taking AI governance seriously' had become a requirement for political credibility in the technology policy domain.

The third dynamic is economic interest. AI systems are deployed across borders. A credit-scoring system developed by a US fintech and deployed by a European bank to assess loan applications from EU residents is subject to the EU AI Act's obligations, regardless of where it was developed. This extraterritorial logic, which the EU AI Act shares with GDPR, means that non-EU organizations face compliance obligations they cannot avoid simply by locating their operations outside EU territory. Other jurisdictions are following similar logic, creating overlapping extraterritorial reaches that affect any globally operating AI developer or deployer. The overlapping extraterritorial effects of multiple binding frameworks are the central compliance management challenge for global enterprises.

Diagram 1.7 - Global AI Regulatory Convergence

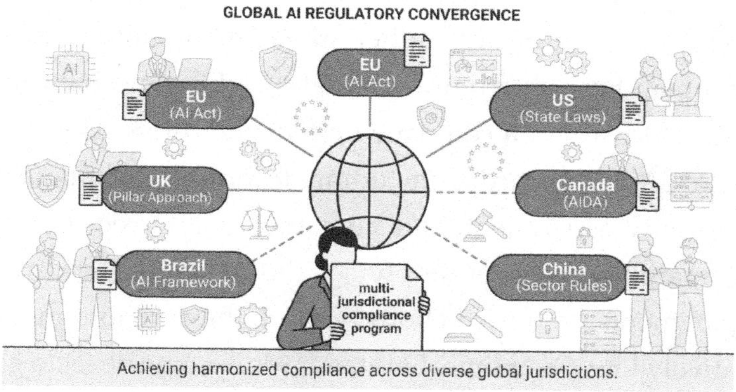

The Influence of the EU's Approach on Global Regulatory Design

The EU AI Act's influence on other jurisdictions' regulatory design is already visible, though it does not take the form of simple replication. The risk-based architecture, which ties compliance burden to potential harm level, has been adopted in modified forms by several jurisdictions, including Canada and, in its sector-specific context, the United Kingdom. The conceptual distinction between providers (who develop and place AI systems on the market) and deployers (who use AI systems in their operations) has influenced how other frameworks allocate responsibility across the supply chain. The requirement for human oversight mechanisms in high-stakes decisions has appeared, in various forms, in US state-level AI laws, NIST's AI Risk Management Framework, and Singapore's voluntary Model AI Governance Framework.

This influence is sometimes described as the 'Brussels Effect' - the tendency for EU regulatory standards to spread globally because organizations seeking EU market access build their products to EU specifications and then apply those same specifications globally rather than maintaining separate versions for different markets. The mechanism is straightforward: if a company builds a high-risk AI system to EU AI Act standards (including technical documentation, conformity assessment, and human oversight requirements), the marginal cost of applying those same standards to deployments in jurisdictions with lighter requirements is relatively low. Over time, EU standards become de facto global minimums for organizations with any EU presence.

For enterprise managers, the Brussels Effect has practical implications: building a compliance program anchored in the EU AI Act's requirements provides a reasonable foundation for multi-jurisdictional compliance, as EU requirements are often among the most comprehensive. This does not mean that EU compliance equals global compliance. Canada's AIDA has requirements that do not map directly onto EU obligations. China's generative AI rules impose obligations that the EU AI Act does not address in the same way. US state laws have specific procedural requirements that differ from EU conformity assessment processes. But a compliance architecture built to EU AI Act standards provides a stronger starting point than one built to any other single jurisdiction's requirements.

Diagram 1.8 - The Brussels Effect on AI Standards

The Brussels Effect on AI Standards

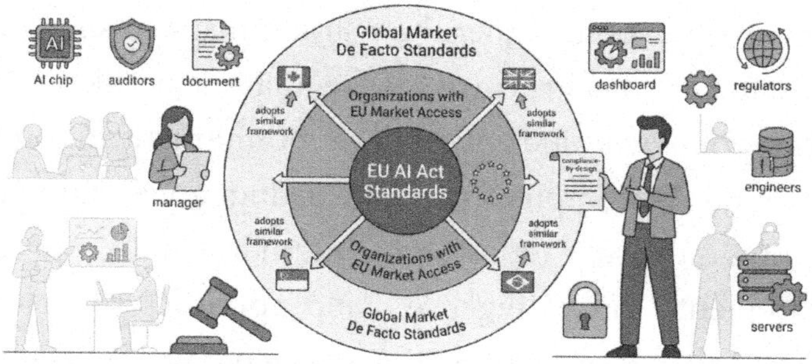

EU AI regulations influence global compliance practices.

What Leadership in AI Governance Means for Enterprise Operating Models

The shift from voluntary principles to binding law has practical consequences for the structure of enterprise operating models. AI compliance is no longer a workstream that can be managed at the level of an individual model-risk team or data-science ethics function. It requires integration across legal, technology, risk, procurement, product development, and human resources functions, because the EU AI Act's obligations touch all of those domains. Technical documentation requirements fall to engineering and data science teams. Conformity assessments require legal and risk expertise. Human oversight mechanisms require HR and operational process design. Post-market monitoring requires data infrastructure and ongoing analytical capacity. No single function owns all of these capabilities.

Organizations that are managing this cross-functional challenge effectively have generally done two things. First, they have assigned explicit ownership at a seniority level sufficient to convene representatives from all relevant functions. A 'head of AI compliance' or equivalent role, reporting at the executive or senior director level, with a remit that spans functions rather than sitting inside a single one, is the organizational structure that regulators are likely to view as evidence of genuine governance commitment. Second, they have built AI compliance into existing governance cycles, including board reporting, internal audit programs, procurement due diligence, and product development gatekeeping, rather than treating it as a parallel process.

The second shift concerns how organizations view AI governance as a leadership responsibility. The EU AI Act creates pathways for personal accountability that did not exist under voluntary frameworks. Senior executives who authorize the deployment of high-risk AI systems are signing off on compliance decisions that can produce substantial regulatory liability. Board members who approve an AI strategy without ensuring that compliance infrastructure is in place face the kind of governance responsibility that courts and regulators will scrutinize in enforcement proceedings. This does not mean that every board member needs to become an AI expert. It means that every board member needs to ask the same questions about AI compliance that they would ask about financial controls: Who is responsible? What are we required to do? How do we know we are doing it? What is the evidence?

Diagram 1.9 - Cross-Functional AI Compliance Ownership

Cross-Functional AI Compliance Ownership

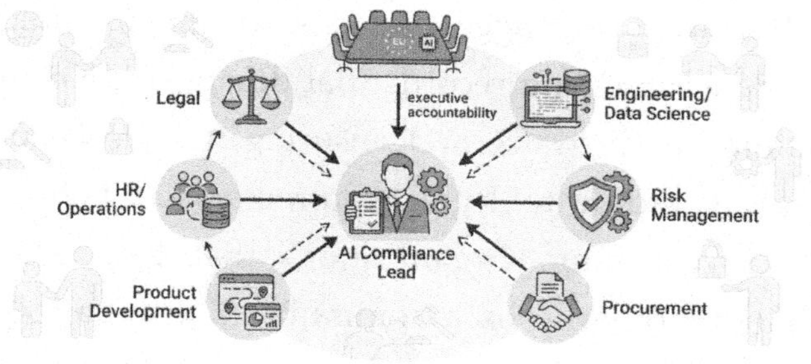

Distributed responsibility ensures holistic AI governance across all business functions.

3.5 What Enforceable Law Changes for Enterprise Leaders

The most important thing a manager needs to understand about the EU AI Act is not any particular provision or definition. It is the structural change that the binding law produces in the relationship between an organization and its regulators. Under a voluntary framework, the relationship was essentially asymmetric in the organization's favor: the organization defined its own compliance, reported on its own performance, and faced limited external scrutiny unless something went visibly wrong. Under a binding law, that asymmetry reverses: supervisory authorities have investigative powers, affected individuals have enforceable rights, and the burden of demonstrating compliance falls on the organization, not on the regulator to prove non-compliance.

The Shift from Reputational Risk to Legal Liability

Reputational risk and legal liability are related but distinct risk categories that require different management approaches. Reputational risk is managed primarily through communication: explaining decisions, demonstrating commitment to values, and responding quickly to criticism. Legal liability is managed through documentation, process, and evidence: creating records that demonstrate compliance, maintaining audit trails, and ensuring that decision-making processes conform to legal requirements, whether or not anyone is watching. Organizations that treat EU AI Act compliance primarily as a communications challenge - 'we need to be able to explain what we are doing' - are addressing reputational risk, not legal liability.

The EU AI Act's penalty structure is designed to make legal liability a serious financial concern for organizations of all sizes. For the most serious violations (deploying prohibited AI practices, failing to correct non-compliant high-risk systems when ordered to do so), fines can reach 35 million euros or seven percent of global annual turnover, whichever is higher. For general non-compliance with high-risk system obligations, the ceiling is 15 million euros or 3% of global turnover, whichever is higher. These are not amounts that can be absorbed as a cost of doing business, and the global turnover basis means that the penalty for a large multinational is calculated on the full enterprise, not just EU revenue. The financial stakes are sufficient to support investment in compliance infrastructure that would have been difficult to justify under a voluntary framework.

For managers, the shift to legal liability also changes the calculus around disclosure and documentation. Under voluntary frameworks, organizations had strong incentives to minimize the paper trail around AI governance decisions, because detailed records of deliberations could document the consideration and rejection of risk concerns in ways that created reputational exposure. Under binding law, the opposite incentive applies: detailed records of governance decisions, risk assessments, and remediation actions are the primary evidence of compliance. A decision log showing that a human oversight mechanism was reviewed, tested, and found adequate is protective. The absence of such a log is itself evidence of non-compliance. This inversion of documentation incentives is one of the most significant behavioral changes that binding law produces.

Diagram 1.10 - Reputational vs. Legal Risk Management

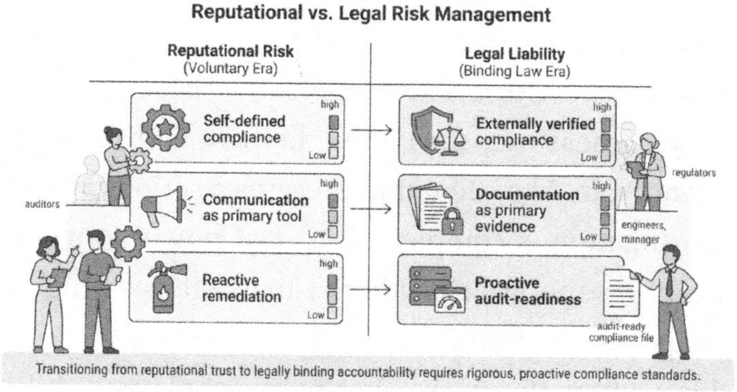

Supervisory Authorities, Market Surveillance, and Enforcement Mechanisms

The EU AI Act creates a layered enforcement architecture. At the European level, the Act establishes the AI Office within the European Commission, which has primary supervisory responsibility for general-purpose AI models and a coordination role across member-state enforcement. At the national level, each member state designates a national competent authority responsible for supervising the Act's implementation within its territory. These national authorities are also market surveillance authorities: they can inspect AI systems, request technical documentation, order testing, and impose corrective measures or penalties. The coordination between national authorities and the AI Office is governed by a consistency mechanism similar to the one used in GDPR enforcement, which means that major cases involving cross-border deployments will involve multiple supervisory bodies.

Market surveillance is the practical mechanism through which enforcement reaches operating companies. Market surveillance authorities can require providers and deployers to provide technical documentation on request, can order independent conformity assessments where they have grounds to believe an AI system poses a risk, and can impose provisional measures (including market withdrawal) when they identify an AI system that presents a serious risk before a full investigation is complete. This last power is particularly significant: a supervisory authority that believes a high-risk AI system poses an immediate risk of harm can require its withdrawal from the market on a provisional basis, shifting the burden to the organization to demonstrate that the system is compliant before it can resume operation.

For compliance program design, the market surveillance architecture has two important implications. First, technical documentation must be maintained in a genuinely audit-ready state, not assembled retrospectively when a request arrives. Supervisory authorities operating under time pressure are unlikely to accept assurances that documentation 'can be produced' within a timeline they consider acceptable. Second, the coordination mechanism means that a single significant case involving one national authority can attract AI Office scrutiny and potentially involve multiple member states' authorities simultaneously. Organizations with EU-wide deployments should assume that a significant enforcement inquiry in one jurisdiction will not be contained to that jurisdiction.

How the Accountability Chain Now Reaches into the Boardroom

One of the most significant structural changes the EU AI Act produces is the extension of accountability from operational AI teams to executive leadership and board level. This shift is not always explicit in the Act's text, which does not routinely name 'board members' or 'CEOs' as directly liable parties. It operates through the accountability chain that the law creates. When an organization is found non-compliant with high-risk AI obligations, regulators will ask how compliance was governed: What oversight structure was in place? Who approved the deployment? What questions did the board ask about compliance before authorizing the AI strategy? The answers to these questions determine whether non-compliance reflects an operational failure or a governance failure, and governance failures attract more severe regulatory and judicial scrutiny.

In practice, boards are now being asked to do for AI compliance what they have done for financial controls and cybersecurity: establish clear ownership, receive regular reporting, and ensure that adequate resources are allocated to compliance infrastructure. The specific questions boards should be asking include: Has the organization completed a comprehensive inventory of its AI systems and their risk classifications under the EU AI Act? Is there a designated AI compliance function with sufficient authority and resources? Has the organization completed conformity assessments for all systems that appear to qualify as high-risk? Are post-market monitoring processes in place and producing regular reports? Has the organization's approach to vendor AI procurement been updated to address EU AI Act obligations that flow through supply chains?

Managers who need to brief boards or executive committees on AI Act compliance are in a better position if they can frame the conversation in terms of governance infrastructure rather than technical detail. A board does not need to understand the specific requirements of Annex IV technical documentation. It needs to determine whether the organization has assigned responsibility for producing that documentation, whether adequate resources have been allocated, whether documentation is being maintained consistently across the AI system portfolio, and whether there is a process for reviewing and updating documentation when systems change. The governance-infrastructure framing translates AI compliance into terms boards can evaluate, yielding the kinds of decisions and records that regulators look for when assessing whether an organization's AI governance is genuine.

3.6 Manager's Checklist

The following items represent the foundational governance actions that enterprise managers should be able to confirm before treating AI Act compliance as a standing operational function rather than an active remediation project.

Each item is verifiable through a document, a process record, or a designated accountability assignment.

- Confirm that your organization has completed a documented inventory of all AI systems in production or in active development, with a preliminary risk classification against EU AI Act categories for each system.
- Verify that a designated AI compliance function or officer exists with a defined reporting line to executive leadership and a remit that spans legal, engineering, risk, procurement, and product functions.
- Confirm that all systems provisionally classified as high-risk have a gap assessment completed against the full set of EU AI Act high-risk obligations, including data governance, technical documentation, human oversight, accuracy and robustness, and post-market monitoring.
- Verify that your board or equivalent governing body has received at least one AI compliance briefing that covers regulatory scope, penalty exposure, and the organization's current compliance status against its highest-risk systems.
- Confirm that AI compliance requirements have been integrated into your standard vendor and procurement due diligence processes, and that AI-specific contractual provisions are in place for third-party AI systems that your organization deploys.
- Verify that technical documentation for each high-risk system is maintained in a format that could be produced to a supervisory authority within a defined

response window, and that the team responsible for maintaining this documentation is identified.

- Confirm that post-market monitoring plans exist for all deployed high-risk systems, specifying the metrics to be tracked, the review frequency, the threshold for escalation, and the process for documenting and reporting serious incidents.
- Verify that your organization has a process for tracking EU AI Act implementation milestones, harmonized standards publications, and national supervisory authority guidance, and that the AI compliance function is reviewing the outputs of that tracking process.

3.7 Takeaway

The EU AI Act marks a permanent shift in the governance environment for AI, not a temporary regulatory cycle that will relax once initial enforcement settles. The forces that drove voluntary principles to binding law - documented failures in consequential AI deployments, structural accountability gaps, and the demonstrated enforceability of GDPR - are durable. They will continue to shape regulatory ambition across jurisdictions long after the Act's initial implementation deadlines have passed. A compliance program built on this understanding is fundamentally different from one built on the assumption that the regulatory environment might soften: it invests in infrastructure, documentation, and organizational capability rather than tactical responses to specific requirements.

For enterprise leaders, the most important single action the EU AI Act demands is assigning clear ownership at an organizational seniority level sufficient to drive cross-functional change. Every other element of compliance - the technical documentation, the conformity assessments, the post-market monitoring, the vendor due diligence - requires coordination across functions that rarely collaborate naturally. The leader who convenes that coordination, maintains its momentum amid competing organizational priorities, and ensures that compliance evidence is produced and consistently maintained is the person who makes the compliance program real. Identifying and empowering that person is the first decision, and the most consequential one.

4 Anatomy of the EU AI Act

A chief technology officer at a global medical device manufacturer received the assignment in straightforward terms: determine which of the company's AI-assisted diagnostic tools are subject to EU AI Act obligations and what those obligations require. What she discovered when her team began the classification work was that the Act's architecture is considerably more layered than a single list of requirements. A radiology AI that identifies abnormalities in mammography scans sits in a different regulatory position than the AI model the company uses to prioritize its own internal quality-control inspections. An AI tool used by hospital purchasing departments to compare suppliers still sits in a different position. Each system required its own classification analysis, and each classification produced a different set of obligations. The exercise took six weeks and involved the legal team, the regulatory affairs function, the engineering group, and external counsel before it produced a result that all parties were confident was defensible.

The CTO's experience highlights the central challenge in EU AI Act compliance: the Act does not impose a uniform set of requirements across all AI systems. It imposes differentiated requirements based on a classification framework, and getting the classification right is the prerequisite for everything else. If a system is incorrectly classified as high-risk, unnecessary compliance resources may be allocated. Conversely, misclassifying a high-risk system as limited-risk can subject the organization to regulatory liability that could have been mitigated through timely investment. Understanding the Act's architecture - its tier structure, its category definitions, its role allocations, and its timeline - is the foundational work that makes every subsequent compliance decision possible.

4.1 Why It Matters

The EU AI Act's risk-tier architecture is not bureaucratic complexity for its own sake. It reflects a deliberate policy judgment that the compliance burden should be proportional to potential harm. A movie recommendation algorithm and a system that evaluates job applications do not pose comparable risks to affected individuals, and requiring the same documentation and oversight mechanisms for both would impose costs that bear no relationship to the actual regulatory concern. The tiered approach allows organizations to concentrate compliance resources on the systems that actually present significant risk, provided that the classification process is rigorous and well-documented.

For managers, this proportionality logic has a direct implication: the Act rewards organizations that invest in systematic classification before they invest in specific compliance measures. An organization that builds its high-risk compliance program before completing a thorough portfolio review risks building it around the wrong systems or failing to include systems that should have been in scope. The classification exercise is not a one-time event. As AI systems are updated, as new systems are deployed, and as supervisory authorities publish interpretive guidance that clarifies classification boundaries, the portfolio review needs to be repeated. Making classification a standing operational function rather than a project is the organizational design choice that creates durable compliance capacity.

The Act also matters at the workflow level because it creates explicit accountability for every role in the AI supply chain: providers (who develop and place systems on the market), deployers (who use systems in their operations), importers (who bring non-EU systems into EU markets), and distributors (who make systems available without substantial modification). Each role carries distinct obligations, and when multiple organizations are involved in a single AI system's lifecycle, the question of who is responsible for which obligations is not always immediately obvious. Getting that question right before a system goes into production is dramatically easier than resolving it after a supervisory inquiry has begun.

4.2 The Risk Tier Framework

The EU AI Act organizes all AI systems into one of four risk tiers, each associated with a distinct level of regulatory obligation. The four tiers are unacceptable risk (prohibited practices), high risk (full compliance obligations), limited risk (transparency requirements), and minimal risk (no specific obligations). The tier structure is the organizing logic of the entire Act, and every compliance decision begins with determining which tier applies to a given system. The architecture is not a spectrum where systems slide gradually between categories: it is a set of defined classifications, each with explicit criteria, that produce categorical differences in compliance requirements.

Unacceptable Risk: Prohibited AI Practices and the Rationale for Absolute Bans

The unacceptable-risk tier contains AI practices that the EU AI Act prohibits absolutely: no conformity assessment, no certification, and no number of additional safeguards can make them lawful for deployment in or affecting EU markets. The prohibited list includes AI systems that use subliminal techniques below the threshold of conscious perception to distort behavior, AI that exploits vulnerabilities of specific groups (including children and people with disabilities) to influence behavior in harmful ways, and AI used by public authorities for social scoring that leads to detrimental treatment. Real-time remote biometric identification systems in publicly accessible spaces are prohibited, with narrow law-enforcement exceptions requiring prior judicial authorization and subject to strict conditions.

The prohibition on real-time biometric identification in public spaces deserves particular attention because it represents the Act's most operationally specific restriction on technology deployment. Law enforcement agencies in EU member states cannot deploy facial recognition systems in public spaces for real-time identification except in circumstances defined by national law for specific serious crime investigations, subject to judicial or independent authorization, and for a limited duration. The exceptions do not extend to general surveillance, pre-crime identification, or persistent monitoring of individuals in public spaces. Organizations in the security technology sector, smart city platforms, and retail analytics have found that products they were actively marketing in the EU market required either significant design changes or withdrawal from the market.

For corporate enterprises (as distinct from law enforcement), the prohibited-practices tier is most significant as a supply-chain concern. An enterprise that purchases or deploys an AI system developed by a third party that incorporates prohibited practices bears responsibility for the violation, even if it was not aware that the system included such functionality. This makes vendor due diligence a non-optional component of AI procurement. Every third-party AI system to be deployed in or affecting EU markets must be assessed against the prohibited-practices list before deployment, and the assessment must be documented in a form that demonstrates the organization's good-faith compliance effort.

Diagram 2.1 - The Four-Tier Risk Architecture

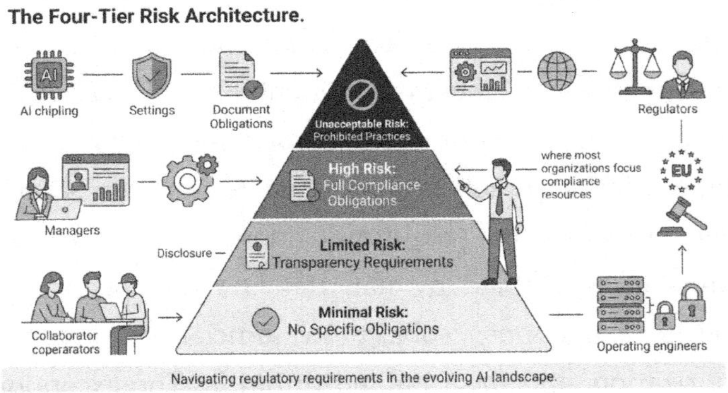

High Risk: The Categories That Trigger Full Compliance Obligations

The high-risk tier is the operational center of the EU AI Act for most enterprise compliance programs. High-risk AI systems fall into two structural categories. The first category covers AI systems that are themselves safety components of regulated products, or are regulated products themselves, under existing EU product safety legislation. This includes AI embedded in medical devices, machinery, aviation components, automotive systems, and other regulated products subject to safety certification under pre-existing EU law. If the AI component is a safety component of such a product, it is a high-risk regardless of any additional analysis.

The second category covers AI systems deployed in specific application domains listed in Annex III of the Act. These domains include biometric identification and categorization, management and operation of critical infrastructure, education and vocational training (including admission, evaluation, and behavioral monitoring), employment and worker management (including recruitment, selection, performance evaluation, and task allocation), access to essential private and public services (including creditworthiness assessment, insurance risk assessment, and emergency response prioritization), law enforcement, migration and border control management, and administration of justice. The breadth of the Annex III list is significant: it covers the majority of consequential AI applications in enterprise and public-sector contexts.

An AI system in an Annex III domain is high-risk if it poses a significant risk of harm to the health, safety, or fundamental rights of natural persons. The Act provides that systems in certain Annex III domains are presumptively high-risk, meaning that the burden of demonstrating that a system does not present such risk falls on the organization deploying it. For most compliance programs, the practical approach is to treat Annex III domain membership as strong evidence of high-risk classification requiring detailed analysis, rather than treating it as a definitive classification requiring no further examination or as an excludable category requiring no attention.

Diagram 2.2 - High-Risk System Classification Paths

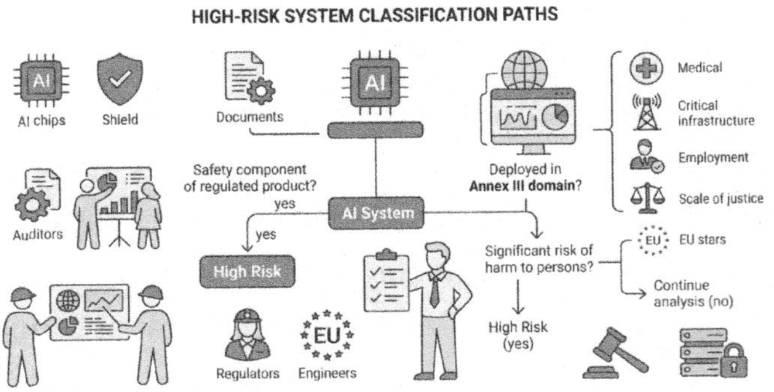

Determining if an AI system falls under the "High Risk" classification based on regulatory frameworks.

Limited and Minimal Risk: Transparency Requirements and the Baseline for Lower-Risk Systems

AI systems in the limited-risk tier are subject to transparency obligations but not the full compliance apparatus that applies to high-risk systems. The primary transparency obligations apply to AI systems that interact directly with natural persons (chatbots and conversational AI), AI systems that generate or manipulate content (deep fakes, AI-generated images, AI-generated text), and AI systems that make individualized recommendations or decisions visible to users. Organizations deploying such systems must ensure that the individuals interacting with or affected by them are informed that they are interacting with AI, unless the AI nature of the interaction is obvious from context.

The minimal-risk tier covers the majority of AI applications by volume, though not by compliance significance. AI used for spam filtering, AI-powered search recommendations, AI-driven logistics optimization, and the many AI tools embedded in standard enterprise software applications generally fall into this tier. The EU AI Act imposes no specific obligations on minimal-risk systems, though it encourages the development of voluntary codes of conduct. Importantly, minimal-risk classification does not exempt a system from other applicable laws: a minimal-risk AI system that processes personal data remains subject to the GDPR. A minimal-risk AI system embedded in a regulated financial product is still subject to applicable financial services law. The EU AI Act tier only determines EU AI Act compliance obligations.

One practical complication in tier classification is that many enterprise AI deployments involve systems that have multiple components or functions, some of which may fall into different tiers. A large enterprise platform might include a high-risk employee performance evaluation module alongside a minimal-risk productivity analytics dashboard, both built on the same underlying model infrastructure. In such cases, the compliance obligations attach to the specific components and their specific use cases, not to the platform as a whole. This disaggregation is necessary for accurate compliance scoping. Still, it adds complexity to the classification exercise, particularly for organizations that have deployed AI through platform integrations where component boundaries are not always clearly defined.

Diagram 2.3 - Limited vs. Minimal Risk Obligations

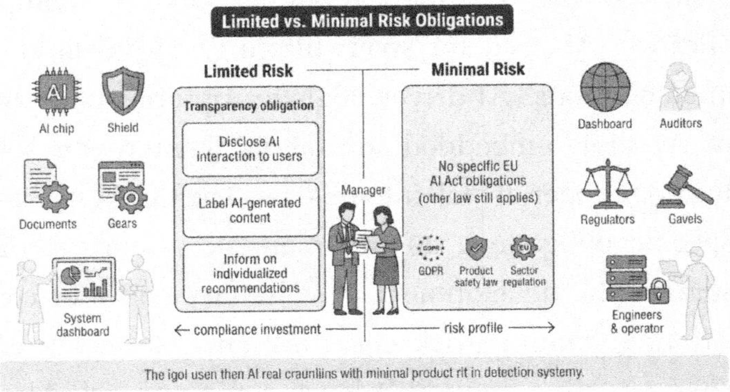

4.3 Prohibited Practices in Detail

The EU AI Act's prohibited practices constitute the regulatory floor: AI applications that the EU has determined pose risks so severe that no business justification or technical safeguard can render them acceptable. In enterprise compliance programs, prohibited practices are not primarily a risk to be managed through mitigation measures. They are practices to identify and eliminate from the portfolio. The organizational challenge is to ensure that the prohibition applies not just to systems the enterprise itself develops, but to every third-party AI system it deploys.

Subliminal Manipulation, Social Scoring, and Real-Time Biometric Surveillance

Subliminal manipulation in the context of the EU AI Act refers to AI systems that use techniques below the threshold of conscious perception to alter an individual's behavior in a way the individual would object to if aware of it, and that cause, or are likely to cause, that individual or another person significant harm. The prohibition targets the combination of covert technique and harmful outcome: it is not a blanket prohibition on persuasive AI or recommendation systems that influence behavior. A content recommendation algorithm that influences viewing behavior by presenting appealing content above the threshold is not subliminal manipulation under the Act's definition. An AI system that uses subliminal audio cues to increase compulsive purchasing behavior in a way that causes financial harm.

Social scoring by public authorities is prohibited in all contexts where AI-generated or AI-assisted assessments of natural persons based on their social behavior or personal characteristics result in detrimental treatment in unrelated social contexts, or in disproportionate or unjustified treatment. The prohibition targets government-operated scoring systems that create comprehensive behavioral profiles to restrict access to services or opportunities. It does not apply to creditworthiness assessments (which are regulated separately as high-risk systems under Annex III), to employee performance assessments used for employment decisions within a single employment relationship, or to security clearance systems governed by national law. But it applies to any public-authority system that generates aggregate social-behavioral scores used to determine treatment across multiple life domains.

Biometric categorization systems that categorize natural persons based on biometric data to infer sensitive attributes, including race, political opinion, trade union membership, religious beliefs, and sexual orientation, are prohibited. This prohibition has particular significance for organizations in the HR technology, insurance, and marketing analytics sectors, where biometric data analysis has been used to infer demographic or behavioral characteristics. The prohibition applies regardless of the stated purpose of the categorization system: it is not sufficient to argue that the system was designed for a different purpose if the operational effect is categorization based on sensitive attributes inferred from biometric data.

Diagram 2.4 - Prohibited AI Practice Categories

The Territorial Scope of Prohibitions and Their Application to Non-EU Developers

The EU AI Act's prohibitions apply to any AI system placed on the EU market or put into service within the EU, regardless of where the system's developer or provider is established. This extraterritorial scope is one of the Act's most significant features for global enterprises and their supply chains. A US-based AI vendor that sells a system incorporating prohibited functionality to a European organization has placed a prohibited system on the EU market. Both the vendor and the

European organization can face regulatory consequences: the vendor for placing a prohibited system on the market, and the organization for deploying it.

The practical compliance challenge this creates is due diligence at the point of procurement. An enterprise deploying a third-party AI system - whether developed by a large platform provider, a specialized AI vendor, or an open-source project - needs to assess that system against the prohibited-practices categories before deployment. This assessment cannot be entirely delegated to the vendor's compliance representations. Supervisory authorities investigating a prohibited-practice violation will look at what due diligence the deploying organization conducted, not just at what the vendor said in a contract. A contractual warranty that a system does not incorporate prohibited practices provides some protection, but it does not substitute for an operational assessment of what the system actually does.

For organizations that deploy AI via API integrations with large platform providers, the prohibited-practices assessment requires attention to the underlying model's capabilities and how those capabilities could be accessed through the API. A language model accessed through an API may be capable of generating outputs that constitute prohibited manipulation if prompted in specific ways. The deploying organization's obligation is not limited to its own prompts and use cases: it extends to ensure that, as deployed, the system cannot be used in prohibited ways by individuals who interact with it. This creates a design obligation that goes beyond vendor selection and extends into how the system is configured, constrained, and monitored in production.

Diagram 2.5 - Extraterritorial Reach of Prohibitions

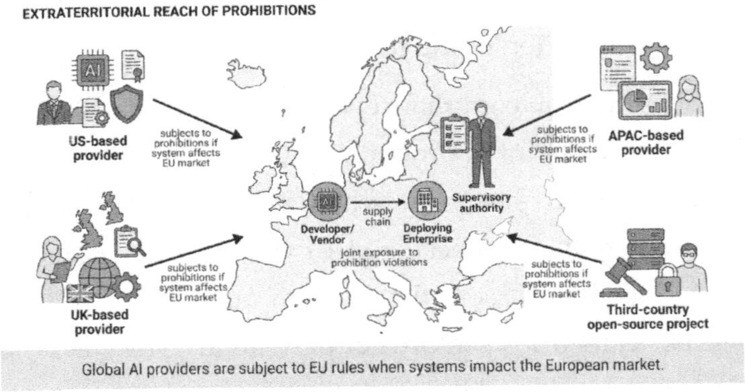

Enforcement Posture and Early Supervisory Guidance on Prohibited Practices

The EU AI Act's prohibited practices took effect in August 2024, making them the earliest-in-force substantive provisions of the Act. The AI Office, established within the European Commission, holds primary supervisory responsibility for prohibited practices at the EU level, with national competent authorities holding concurrent enforcement authority at the member-state level. Early supervisory statements have indicated that the AI Office intends to use its market surveillance powers proactively: it has signaled interest in investigating systems in the biometric identification and categorization space, and in monitoring the behavior of large general-purpose AI providers whose systems might enable prohibited uses when deployed by third parties.

For compliance program design, the early enforcement focus on prohibited practices has several implications. First, prohibited-practice screening should be prioritized before the organization invests in the more complex conformity-assessment work required for high-risk systems. A prohibited-practice violation is categorically more serious than a high-risk compliance gap: it cannot be remediated by improving documentation or adding an oversight mechanism. It can only be remediated by removing the system from the market or redesigning it to eliminate prohibited functionality. Second, the due diligence trail for prohibited-practice screening should be treated as a permanent compliance record, not a project artifact. If a supervisory authority investigates a deployed system years after procurement, the organization's ability to demonstrate that it conducted appropriate due diligence at the time of procurement will depend on records maintained from the beginning.

Organizations in the recruitment technology, HR analytics, and customer behavior analytics sectors face the highest immediate exposure to prohibited-practices enforcement because these sectors are where biometric categorization, manipulation-adjacent techniques, and vulnerability exploitation are most likely to appear in commercial AI products. A global logistics company deploying an AI workforce management system that uses physiological signals to assess worker productivity should treat that system's compliance with the prohibited-practices provisions as an urgent priority, not a future compliance milestone. The prohibited practices are not a future concern with a distant effective date: they are current law and a current enforcement priority.

Diagram 2.6 - Prohibited Practices Enforcement Timeline

Prohibited Practices Enforcement Timeline

Key compliance dates for EU AI Act enforcement regarding banned applications.

4.4 High-Risk System Obligations

High-risk AI systems carry the most comprehensive compliance obligations in the EU AI Act. These obligations are not a checklist of administrative formalities: they are substantive requirements that extend to the design, documentation, deployment, and monitoring of systems. Meeting them requires sustained investment in technical and organizational capability, and the timeline for building that capability is shorter than many organizations initially estimate.

The Annex Classification Lists and How to Apply Them to Your Product Portfolio

Annex III of the EU AI Act lists the high-risk application domains for the second classification category (systems not already captured as safety components of regulated products). The Annex is structured around eight domain areas, each with specific descriptions of the types of AI applications that fall within scope. Applying the Annex to a real product portfolio requires close reading of those descriptions, because the scope boundaries are not always immediately obvious from the domain labels. 'Employment and worker management,' for example, covers not just recruitment and selection tools but also systems for task allocation, monitoring workers' performance, and evaluating employees throughout the employment relationship. A performance dashboard used by a call center supervisor to track agent productivity metrics may fall within scope depending on how the AI-generated assessments feed into employment decisions.

The classification exercise requires both legal interpretation and operational knowledge. Legal teams typically cannot determine whether a system falls within the scope of Annex III without understanding what the system actually does: what inputs it uses, what outputs it produces, what decisions those outputs inform, and how those decisions affect individuals. This means the classification exercise must involve both legal and technical stakeholders, and the exercise record must document both the legal analysis and the factual basis for the conclusions reached. A classification decision based on a general description of the system's purpose rather than a specific review of its operational behavior is not defensible if challenged by a supervisory authority.

Organizations with large AI portfolios should approach the Annex III classification exercise systematically rather than system by system in isolation. A master classification framework that maps system types to Annex III domains, with documented legal analysis for each mapping, enables consistent application of the classification work across the portfolio and efficient updates when regulatory guidance clarifies interpretation questions. The AI Office and national competent authorities are expected to publish interpretive guidance on Annex III boundaries as implementation proceeds: organizations that have built a documented classification framework can update their analyses in response to that guidance without starting from scratch, while organizations that have not built such a framework face the risk of having made classification decisions they cannot efficiently review.

Diagram 2.7 - Annex III Domain Classification Map

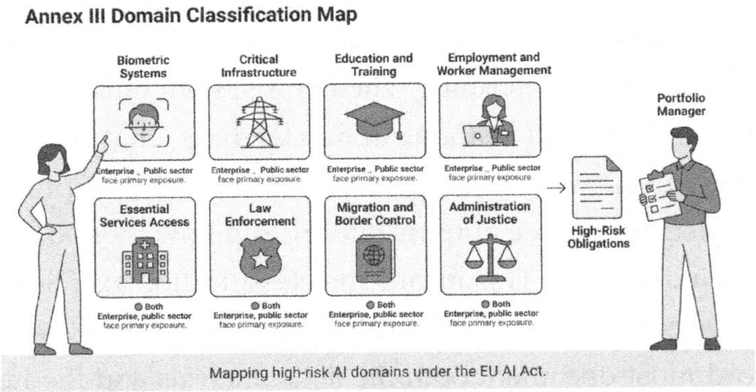

Core Obligations: Data Governance, Documentation, Transparency, Human Oversight

High-risk AI systems must satisfy six substantive obligation categories, each of which requires organizational capability to implement and maintain. The first is data governance: training, validation, and testing data must meet quality criteria relevant to the system's intended purpose, must be processed in accordance with appropriate data practices for the application domain, and must be free from practices that produce discriminatory outputs in the context of the system's intended use. Data governance for AI is not the same as data governance for data warehousing or reporting: it requires attention to training data composition, labeling quality, representation of relevant demographic groups in test sets, and ongoing monitoring of data quality as it shifts over time.

Technical documentation under Annex IV is the second major obligation. Annex IV specifies a detailed list of information that must be documented before a high-risk system is placed on the market or put into service. This includes a general description of the system and its intended purpose, the design specifications, the training methodology and data characteristics, the performance metrics and testing results, the human oversight mechanisms, the cybersecurity measures, and the post-market monitoring plan. This documentation must be prepared before the system goes into production, maintained and updated throughout its operational life, and made available to supervisory authorities on request. The production of Annex IV documentation is a significant technical and organizational undertaking, particularly for organizations that have not historically maintained model documentation at this level of detail.

Human oversight is the third category, and it is often the most operationally complex. High-risk AI systems must be designed and deployed in a way that allows natural persons to effectively oversee the system, understand its outputs, intervene or override its decisions, and shut it down in the event of malfunction. This is not a requirement to have a human sign off on every AI-generated output; it is a requirement to design the operational context so that human oversight is genuinely possible and effective. A human who reviews an AI-generated credit score without having access to the features that drove the score, the confidence level of the prediction, and a mechanism to override the score is nominally in the oversight loop but is not exercising meaningful oversight. The Act requires meaningful oversight, and supervisory authorities are likely to scrutinize oversight mechanisms specifically during enforcement reviews.

Diagram 2.8 - High-Risk System Obligation Framework

The Provider-Deployer Distinction and How Responsibility Is Allocated Across the Supply Chain

The EU AI Act creates four distinct roles in the AI supply chain, each with different legal obligations: providers, deployers, importers, and distributors. Understanding your organization's role in a given AI system determines what you are required to do. Providers are the organizations that develop high-risk AI systems and place them on the market under their own name or trademark, or have a high-risk system developed to their specification. Providers bear the most comprehensive obligations: conformity assessment, Annex IV technical documentation, registration in the EU database for high-risk systems, CE marking (where applicable), and a post-market monitoring system.

Deployers are organizations that use high-risk AI systems under their own authority, other than for personal use. Deployers are typically the businesses that integrate a provider's AI system into their own operations and workflows. A bank that purchases a credit-scoring AI from a vendor and uses it to assess loan applications is a deployer. Deployers have their own set of obligations: ensuring that AI is used in accordance with its intended purpose and the provider's instructions, implementing human oversight measures as specified by the provider, monitoring the AI's operational performance, informing and training staff who work alongside the system, and informing supervisory authorities and affected individuals of serious incidents. Deployers do not bear the full conformity-assessment burden of providers, but they are not passive recipients of compliance; they bear active obligations in the deployment context.

The interaction between provider and deployer obligations is most complex when a deployer makes significant changes to a provided AI system or deploys it for a purpose different from the provider's intended use. In such cases, the deployer may become a provider for regulatory purposes and take on the full set of provider obligations. This means that organizations that routinely fine-tune, retrain, or adapt third-party AI systems to their specific operational context need to assess whether those adaptations are substantial enough to warrant a shift in their regulatory role. The analysis is not binary: it depends on the nature and scope of the modification, the modified system's risk profile, and the extent to which the modification alters the system's compliance characteristics. Legal and technical teams need to conduct this analysis jointly, and the conclusion needs to be documented as part of the organization's compliance record.

Diagram 2.9 - AI Supply Chain Role Obligations

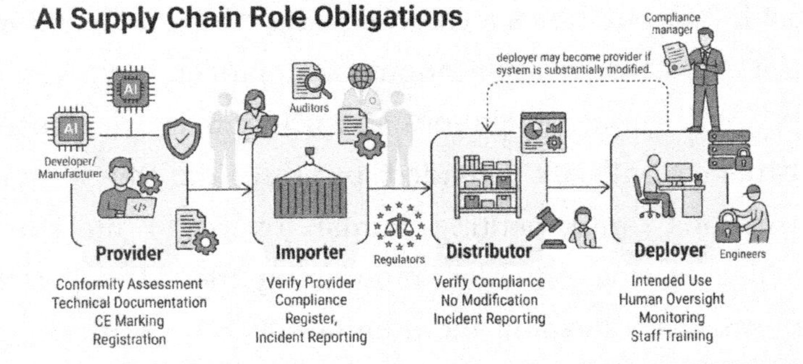

Understanding responsibilities at each stage of the AI supply chain under regulations.

4.5 General-Purpose AI: A New Regulatory Category

One of the EU AI Act's most significant innovations is the creation of a distinct regulatory category for general-purpose AI (GPAI) models: large AI systems trained on broad datasets that can perform a wide range of tasks and integrated into other AI systems or applications. This category was not in the Commission's original 2021 proposal. It was added during the legislative process in response to the rapid emergence of large language models and foundation models as foundational infrastructure for AI deployment across sectors. The GPAI provisions represent the Act's attempt to address AI systems that do not fit neatly into the provider-of-a-specific-system framework that governs the rest of the regulation.

What Qualifies as a General-Purpose AI Model Under the Act's Definitions

A general-purpose AI model is defined in the Act as an AI model that is trained on a large amount of data using self-supervision at scale, that demonstrates significant generality, and that can competently perform a wide range of distinct tasks and be integrated into a variety of downstream systems or applications. The definition is calibrated to capture foundation models and large language models without requiring that they be available as standalone products: a GPAI model that is only accessible through an API and is always accessed as a component of a larger system still qualifies if it meets the generality and capability thresholds.

The practical scope of the GPAI category is still being refined through AI Office guidance, but the core cases are relatively clear. The major commercial large language models, multimodal foundation models, and large-scale image- and code-generation models qualify. Smaller, purpose-built models trained on narrow data sets for specific tasks generally do not qualify, even if they are technically sophisticated. The distinction matters because GPAI obligations apply specifically to the models that serve as infrastructure for multiple downstream applications, not to every AI system deployed in an organization. An enterprise that uses a commercial GPAI API to power its customer service chatbot needs to assess both the GPAI provider's compliance (relevant to the vendor relationship) and its own deployer obligations (relevant to how the chatbot is configured and operated).

One complication for enterprise compliance programs is the treatment of open-source GPAI models. The Act provides more limited transparency obligations for GPAI models released under open-source licenses, reflecting the policy judgment that open-source publication of model weights provides a form of transparency that commercial closed-source models do not. However, this accommodation does not apply to open-source GPAI models designated as posing systemic risk. Organizations that deploy open-source GPAI models, including those that fine-tune and self-host foundation models, need to assess their position carefully: they may be acting as GPAI providers for regulatory purposes, depending on the scale of their deployment and the extent to which they are making the model available to others.

Diagram 2.10 - GPAI Model Regulatory Scope

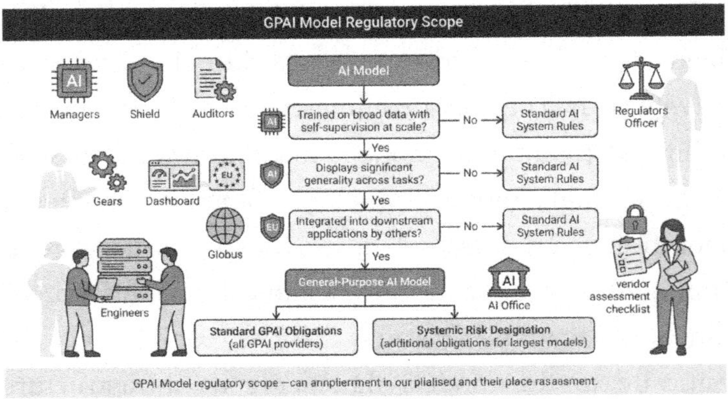

Transparency Obligations for All GPAI Providers

All providers of general-purpose AI models bear transparency obligations under the Act, regardless of whether the model is commercially distributed or open-source. These obligations focus on providing downstream integrators with the information they need to comply with their own obligations when they build applications using the GPAI model. The Act requires GPAI providers to publish technical documentation covering the model's architecture, training data sources and methodologies, capabilities and limitations, performance benchmarks, and measures taken to address known risks. This documentation serves a dual purpose: it enables regulators and researchers to assess the model's properties, and it enables downstream deployers to assess whether the model is suitable for their intended application and what additional safeguards may be needed.

For enterprise managers who are downstream users of GPAI models rather than providers, the transparency obligations create a right to information that supports their own compliance work. When evaluating GPAI vendors, organizations should request and review the technical documentation that the Act requires providers to produce. A GPAI provider that cannot or will not produce compliant technical documentation is itself in violation of the Act's requirements, and an enterprise deploying that provider's model in the EU market inherits the risk of that upstream non-compliance. Vendor due diligence for GPAI-based systems should include explicit review of the provider's GPAI compliance documentation as a procurement gate.

The transparency obligations for GPAI providers also include copyright compliance requirements: providers must implement a policy to comply with EU copyright law, including the text and data mining exceptions. They must publish a sufficiently detailed summary of the content used for training. This requirement is particularly significant for organizations that train GPAI models on proprietary data or use them in content generation workflows. The copyright compliance dimension of GPAI governance is an area where the Act's requirements intersect with ongoing EU and national-level copyright litigation, and where the regulatory landscape is still evolving. Monitoring AI Office guidance on GPAI copyright compliance is a standing action for enterprises active in content generation, media, and publishing.

Systemic Risk Designation and the Heightened Obligations It Triggers

The EU AI Act creates a special designation for GPAI models that present systemic risk, defined by reference to aggregate training compute above a defined threshold (currently set at 10^25 floating-point operations) or by an AI Office determination that the model presents systemic risk based on its reach, capabilities, or the nature of the foreseeable risks it creates. Models with systemic risk designation face significantly heightened obligations compared to standard GPAI providers. These include adversarial testing (red-teaming) of the model before market release and after significant updates, incident reporting to the AI Office for serious incidents caused by the model or its downstream applications, and ongoing assessment and mitigation of systemic risks identified through testing and monitoring.

The systemic risk designation is currently relevant primarily for the largest commercial AI providers: the organizations training foundation models at the frontier of scale and capability. For most enterprise organizations, the immediate practical relevance is as a procurement and vendor-management concern rather than a direct compliance obligation. Enterprises that build significant operations on GPAI models with a systemic risk designation need to actively monitor those models' compliance status, because disruptions to the models' EU market access (due to regulatory action) would have material operational consequences. The contractual relationship with systemic-risk GPAI providers should address what happens to the enterprise's access and operations if the provider becomes subject to corrective measures or market access restrictions.

The systemic risk framework also illustrates the EU AI Act's approach to regulatory evolution. The Act explicitly provides that the AI Office can adjust the compute threshold for systemic risk designation as technology evolves, and that additional criteria can be applied based on qualitative assessment. This means the set of models subject to systemic risk obligations will change over time, and enterprises should monitor AI Office guidance on systemic risk designations as part of their standing regulatory monitoring program. A model that is not currently systemically designated may become so as the AI Office develops its assessment methodology and as model capabilities evolve.

4.6 Manager's Checklist

The following items represent the verification actions that confirm an organization has completed the foundational classification and compliance architecture work required to manage EU AI Act obligations on an ongoing basis. Each item is verifiable through a specific document or organizational structure.

- Confirm that your organization has completed a full AI system inventory that maps every AI system in production or active development to the EU AI Act tier classification (prohibited, high-risk, limited-risk, minimal-risk), with documented legal and factual basis for each classification.

- Verify that all systems provisionally classified as high-risk have been assessed against the complete Annex III domain list and, where applicable, the safety-component-of-regulated-product classification path, with classification conclusions reviewed by legal counsel.
- Confirm that Annex IV technical documentation has been initiated for each high-risk AI system, with a named owner responsible for maintaining and updating the documentation throughout the system's operational life.
- Verify that your organization has assessed its regulatory role (provider, deployer, importer, or distributor) for each AI system in its portfolio, including third-party AI systems accessed through APIs, licensed software, or vendor contracts.
- Confirm that human oversight mechanisms for each high-risk AI system have been designed and documented, specifying who exercises oversight, what information they have access to, how they can intervene or override AI-generated outputs, and what training they receive.
- Verify that post-market monitoring plans are in place for all deployed high-risk systems, including defined metrics, review frequency, escalation thresholds, and the process for documenting and reporting serious incidents to national competent authorities.
- Confirm that your procurement and vendor due diligence process includes prohibited-practices screening for all third-party AI systems before

deployment, with screening results documented in the vendor file.

- Verify that GPAI vendors supplying models used in your EU-market AI applications have been assessed for compliance with their own EU AI Act obligations. That compliant technical documentation has been received and reviewed.
- Confirm that your organization has identified the EU AI Act implementation milestones most relevant to your portfolio (including the August 2026 deadline for most high-risk system obligations) and that a compliance readiness timeline has been established with named milestones and owners.

4.7 Takeaway

The EU AI Act's architecture is the compliance manager's primary orientation tool. Every resource allocation decision, every vendor assessment, every documentation project, and every governance structure traces back to the question of where a given AI system sits in the risk-tier framework. Organizations that invest in getting the classification right - systematically, with documented legal analysis, and with regular review as systems and guidance evolve - build a foundation that makes every subsequent compliance action more efficient and more defensible. Organizations that treat classification as a rough approximation and move quickly to specific compliance measures risk building compliance programs around the wrong scope.

The provider-deployer distinction is the second structural insight that managers need to internalize. Compliance obligations follow roles, not just systems, and role determination requires analysis of what your organization actually does with an AI system, not just what the vendor says about it. An enterprise that customizes, fine-tunes, or substantially configures a third-party AI system may be acting as a provider for regulatory purposes, with all of the compliance obligations that entail. Clarifying this before deployment, documenting the analysis, and ensuring that contractual arrangements with AI vendors accurately reflect the actual risk allocation between the parties are governance actions that create durable compliance capacity in an increasingly complex supply chain.

5 The World Beyond Article One

The global head of AI compliance at a multinational industrial equipment manufacturer had a straightforward question when she joined the company: which jurisdictions require what, and by when? Eighteen months later, she had a more precise understanding of the problem. The company's AI systems span predictive maintenance models deployed at customer facilities across four continents, an HR analytics platform used to inform promotion decisions for employees in 14 countries, and a computer-vision quality-control system used in manufacturing plants in Germany, Mexico, and China. The EU AI Act covers the HR platform and the computer vision system as high-risk deployments. Colorado's AI Act, effective June 2026, covers the HR platform's use in consequential employment decisions affecting Colorado residents. China's algorithmic recommendation and generative AI regulations impose transparency and security-assessment obligations on systems the company uses to communicate with Chinese customers. Canada's Artificial Intelligence and Data Act will eventually impose impact assessment obligations on the HR platform's use in Canada, with enforcement timelines still being finalized as the company's legal team models out its compliance roadmap.

The compliance head's challenge is not that any single framework is unmanageable. Each framework has its own definitions, classification triggers, documentation requirements, and enforcement calendar, and the frameworks do not map neatly onto one another. An impact assessment produced to satisfy the EU AI Act's high-risk documentation requirements shares conceptual overlap with what Colorado's algorithmic impact assessment requires, but the structural differences are sufficient that neither document satisfies the other's requirements without modification. A compliance architecture designed around a single framework requires retrofitting to satisfy other frameworks. Her conclusion after eighteen months: the only sustainable approach is to build a unified compliance architecture that captures the highest-common-denominator requirements across all applicable frameworks, with jurisdiction-specific supplements where required. This chapter explains what she is building and why she is building it.

5.1 Why It Matters

For managers with global AI responsibilities, the most important insight about the international AI regulatory landscape is that it is simultaneously converging and diverging. It is converging on a set of common conceptual commitments: risk-based classification, transparency obligations, human oversight in high-stakes decisions, accountability for AI supply chains, and some form of impact assessment for consequential AI applications. It diverges on the specifics of how those commitments are operationalized, the definitional scope they apply to, the enforcement mechanisms that back them, and what the political theory of AI governance is supposed to accomplish. Managing this convergence-with-divergence requires a compliance architecture that is principled enough to capture the common core and flexible enough to accommodate jurisdiction-specific variation.

The jurisdictional complexity also has a temporal dimension that managers must actively manage. Regulatory frameworks for AI are not static. The EU AI Act's implementation is rolling out across multiple deadline tranches through 2027. Colorado's AI Act takes effect in 2026, with amendments still being debated at the time of its passage. Canada's AIDA is still moving through the legislative process. China's regulatory approach is evolving through successive waves of sector-specific rules. Singapore's AI governance framework is expanding from a voluntary model to a more structured set of expectations. The regulatory landscape a compliance team must navigate in 2026 is materially different from the landscape in 2024, and the landscape in 2028 will be different again. Building a compliance program designed only for the current state of the landscape is a strategy that requires constant retrofitting.

At the workflow level, jurisdictional complexity creates specific management challenges around documentation, vendor selection, product design, and organizational structure. A product team designing an AI-assisted HR system that will be deployed across Europe, North America, and Asia must make design decisions early in the development process that affect compliance across all applicable jurisdictions: what data is collected, what decisions the system informs, what disclosures are made to individuals, and what override mechanisms are available. Making these decisions, jurisdiction by jurisdiction, after the product is built, is dramatically more expensive than building compliance into the product architecture from the start. The pilot-to-production pathway for globally deployed AI systems now requires compliance integration at the design stage, not as a post-production approval step.

5.2 The United States: Fragmentation as a Compliance Environment

The United States does not have a federal AI law. What it has is a proliferating landscape of state-level legislation, sector-specific federal guidance, executive orders with limited binding authority, and agency-level rules that apply AI governance principles within existing regulatory mandates. For enterprise compliance managers, this means the US regulatory environment requires monitoring across multiple legislative and regulatory bodies simultaneously, with no single framework providing uniform coverage across all AI applications and all states. The absence of federal law is itself a compliance strategy consideration: it means that organizations operating in multiple US states face different requirements in different jurisdictions, and the mosaic of state laws is growing.

Colorado's AI Act: Scope, Algorithmic Impact Assessments, and Enforcement

Colorado's Artificial Intelligence Act, signed into law in May 2024 and effective February 1, 2026, is the most architecturally complete US state AI law currently on the books. It applies to developers and deployers of 'high-risk AI systems,' defined as systems that make or substantially inform consequential decisions in domains including education, employment, financial services, healthcare, housing, insurance, and legal services. The consequential-decision trigger is broad: a system that 'substantially influences' a consequential decision falls within scope even if a human makes the final decision. This means AI systems used in loan underwriting, benefits determination, hiring, and insurance risk assessment that affect Colorado residents are likely subject to the Act, regardless of the deploying organization's headquarters.

The Act's core compliance mechanism is the algorithmic impact assessment (AIA), which deployers must complete before deploying a high-risk AI system and on an ongoing annual basis. The AIA requires an evaluation of the system's potential for algorithmic discrimination, an explanation of how the system functions and what safeguards are in place, and documentation of the steps taken to mitigate identified risks. Deployers must also provide Colorado residents with notice that an AI system was used in a consequential decision, the principal reason for the decision, and a means to correct inaccurate personal data. These rights are operationally significant: they require that deployers have the technical capability to explain AI-generated decisions at the individual level and to accept and process correction requests, capabilities that many current AI deployments lack.

Enforcement authority rests with the Colorado Attorney General, who can investigate complaints, compel production of AIAs and related documentation, and bring enforcement actions. The Act also establishes an affirmative defense for deployers who can demonstrate that they followed the developer's instructions for use and that the developer is primarily responsible for the violation, thereby creating an important compliance incentive to ensure that vendor contracts include clear allocations of responsibility between the developer and the deployer. For organizations that deploy third-party AI systems in Colorado, the affirmative defense is only available if the organization actually received and followed the developer's instructions, making the documentation of developer guidance a compliance record of practical legal significance.

Diagram 3.1 - Colorado AI Act Compliance Flow

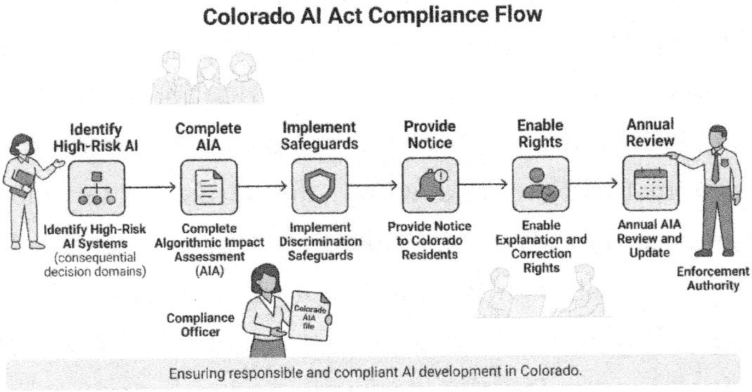

Illinois, California, and the Emerging State Law Landscape

Colorado is the leading example of comprehensive state AI legislation, but it is not the only US jurisdiction that enterprise compliance programs must track. Illinois has a dense cluster of AI-specific provisions across multiple statutes. The Artificial Intelligence Video Interview Act requires employers to notify job candidates when AI is used to analyze video interviews, obtain consent before using AI analysis, and provide candidates with the ability to opt out of AI-based video screening. The Illinois Biometric Information Privacy Act (BIPA), though predating the current AI regulatory wave, imposes strict consent, retention, and data destruction requirements on any organization that collects biometric identifiers, including facial scans and voiceprints, making it one of the most financially consequential AI-adjacent laws in the United States based on the volume of private litigation it has generated.

California has pursued AI governance through multiple legislative channels simultaneously. The California Consumer Privacy Act and its amendment, the California Privacy Rights Act, give California residents rights over automated decision-making that includes explanation and opt-out rights in contexts involving 'significant decisions' - a scope that overlaps substantially with the kinds of AI applications that Colorado and the EU treat as high-risk. California has also enacted specific legislation on AI transparency in political advertising, AI use in healthcare settings, and requirements for AI systems to disclose their AI nature to users. Multiple additional AI-specific bills have been introduced in successive legislative sessions, creating a moving compliance target that California-facing enterprises must track actively.

New York City's Local Law 144, which requires annual bias audits and public disclosure of results for automated employment decision tools used in hiring and promotion decisions affecting NYC residents, represents a distinct regulatory approach: city-level AI-specific compliance obligations that apply to a single, high-stakes use case. The NYC bias audit requirement has generated a market for third-party AI auditors and produced the first substantial body of publicly disclosed AI bias audit results in the United States, creating a de facto performance benchmark that employers using similar tools face pressure to meet. For national and global enterprises with significant workforce presence in New

York City, Local Law 144 compliance is an operational requirement with immediate practical consequences, not a future regulatory risk.

Diagram 3.2 - US State AI Law Landscape

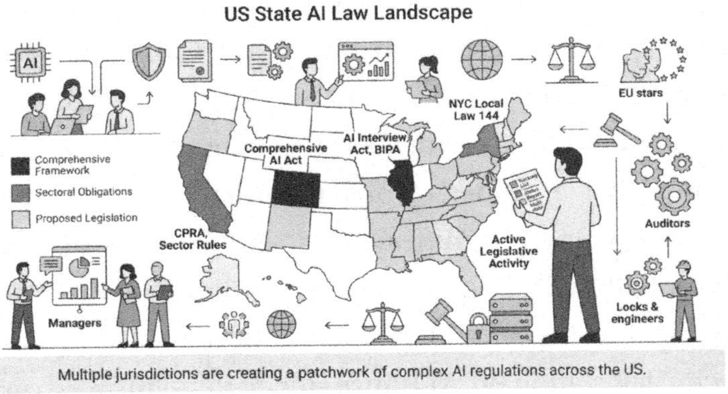

The Absence of Federal AI Law and Its Implications for Enterprise Compliance Strategy

The absence of a comprehensive federal AI law in the United States is not a static feature of the regulatory landscape. It is the product of ongoing political dynamics that could result in federal legislation at any time, and the nature of any eventual federal framework would have significant implications for the structure of enterprise compliance programs. The trajectory has moved from an Obama-era report on algorithmic accountability, through Trump-era executive orders emphasizing AI competitiveness and voluntary principles, through Biden-era executive orders establishing more detailed AI governance expectations for federal agencies and imposing reporting requirements on developers of large foundation models, to an ongoing legislative debate in which multiple comprehensive federal AI bills have been introduced without advancing to final passage.

In the interim, the primary federal AI governance mechanisms are sector-specific regulatory guidance from agencies operating within their existing statutory mandates. The Equal Employment Opportunity Commission has published guidance on AI tools in hiring that applies existing anti-discrimination law to algorithmic systems. The Consumer Financial Protection Bureau has clarified that AI-generated credit decisions must comply with existing fair lending law, including explanation requirements. The Food and Drug Administration has a software-as-a-medical-device framework that addresses AI in medical diagnostics. The Federal Trade Commission has pursued enforcement actions against companies making misleading claims about AI capabilities. These sector-specific mechanisms create a patchwork of federal AI governance obligations that apply to specific industries without providing the comprehensive cross-sector framework that the EU AI Act provides.

For an enterprise compliance strategy, the absence of a comprehensive federal AI law has two principal implications. First, the state-law compliance burden is not a temporary condition that will be resolved when federal legislation preempts state law. Federal preemption of state AI law is not guaranteed, and even if federal legislation passes, it may set a floor rather than a ceiling, leaving state laws with stricter requirements in place. Organizations should build state-law compliance capacity as a standing function, not as a bridge to a simpler post-federal-law environment. Second, the sector-specific federal guidance that does exist is binding to the extent it interprets existing statutes, and organizations that focus exclusively on new AI-specific legislation may be missing compliance obligations that already apply to them under existing law as interpreted by regulatory agencies.

Diagram 3.3 - US Federal AI Governance Landscape

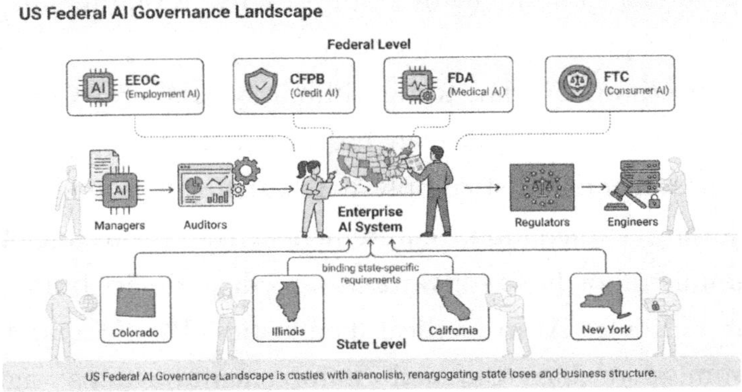

5.3 Canada and the AIDA Framework

Canada's approach to AI regulation is anchored in the Artificial Intelligence and Data Act (AIDA), which was introduced as Part 3 of Bill C-27 in 2022. AIDA represents Canada's first attempt at comprehensive, horizontal AI legislation, designed to apply across sectors rather than targeting specific industries. As of this writing, AIDA has not yet received Royal Assent, and the timeline for its entry into force remains uncertain. However, the substantive framework established by AIDA has been sufficiently detailed and debated that organizations with Canadian operations should build compliance capacity now rather than wait for final enactment. The direction of Canadian AI policy is clear, even if the specific timeline is not.

The Artificial Intelligence and Data Act: Scope and High-Impact System Obligations

AIDA applies to organizations that design, develop, make available for use, or manage the operation of AI systems in the course of international or interprovincial trade and commerce. The use of trade and commerce as the jurisdictional hook reflects Canada's constitutional structure, in which federal legislation in this domain requires a federal head of power. The scope covers the AI supply chain comprehensively: developers who create AI systems, providers who make them available, and operators who deploy them in business contexts. Each role carries distinct obligations under the framework, parallel to the EU AI Act's provider-deployer distinction, though not identical in content.

The central compliance mechanism in AIDA is the requirement to identify, assess, and mitigate risks for 'high-impact AI systems.' A high-impact AI system poses a significant risk of harm to individuals, the environment, or Canadian interests, based on factors such as the nature of the decision, the scale of use, the vulnerability of affected individuals, and the degree of human oversight. The Canadian government has signaled that high-impact systems will likely include AI used in employment decisions, credit and financial decisions, healthcare diagnostics, criminal justice applications, and critical infrastructure management, overlapping substantially with the EU AI Act's high-risk categories but not identical to them.

Operators of high-impact AI systems under AIDA will be required to implement mitigation measures, maintain records of those measures, and make a publicly available description of the system's use. In cases where a high-impact AI system causes or could cause serious harm, operators face additional obligations, including notifying a government minister. Certain violations of AIDA's prohibitions (including making an AI system available when it was not assessed for high-impact status, and failing to implement mitigation measures for identified risks) carry criminal penalties, a significant escalation beyond the administrative fine structures that characterize most AI regulatory frameworks globally.

Diagram 3.4 - AIDA High-Impact System Framework

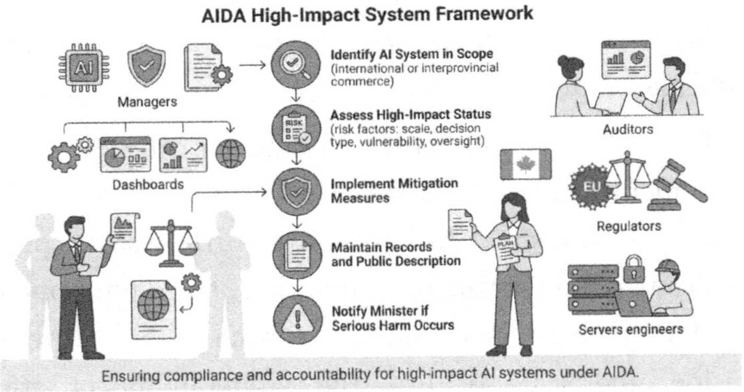

How AIDA Aligns With and Diverges From the EU AI Act

AIDA and the EU AI Act share a common conceptual architecture: both are risk-based, horizontal frameworks that apply across sectors, both create tiered obligations based on a system's potential for harm, and both impose obligations on multiple roles in the AI supply chain. This shared architecture reflects that Canadian regulators closely followed the EU legislative process and drew on its analytical framework in designing AIDA. For organizations that have already built compliance infrastructure around the EU AI Act, this conceptual overlap provides a more useful foundation for AIDA compliance than starting from scratch.

The divergences between AIDA and the EU AI Act are nonetheless significant. AIDA's high-impact classification criteria are less precisely defined than the EU AI Act's Annex III list, giving regulators more discretion in determining scope but also creating more uncertainty for organizations trying to determine whether their systems are in scope. AIDA's mitigation requirements are framed in outcome terms (implement measures that mitigate risk) rather than specifying the detailed content of documentation, testing, and oversight mechanisms that the EU AI Act requires. This gives organizations more flexibility in how they demonstrate compliance, but also creates more room for regulatory disagreement about whether what an organization has done is sufficient.

The most significant structural divergence is in the governance mechanism. The EU AI Act relies on designated national competent authorities in each member state, coordinated at the EU level by the AI Office. AIDA relies on a federal AI and Data Commissioner, a new position created specifically to oversee AIDA implementation. The Commissioner's investigative and enforcement powers, the process for handling complaints, and the relationship between AIDA compliance and compliance with sector-specific Canadian regulations (including the financial sector regulator OSFI's guidance on model risk management) are areas where implementation details will materially affect what compliance looks like in practice. Organizations should monitor the AI and Data Commissioner's early publications closely for the interpretive guidance that will fill these gaps.

Diagram 3.5 - AIDA vs. EU AI Act: Alignment and Divergence

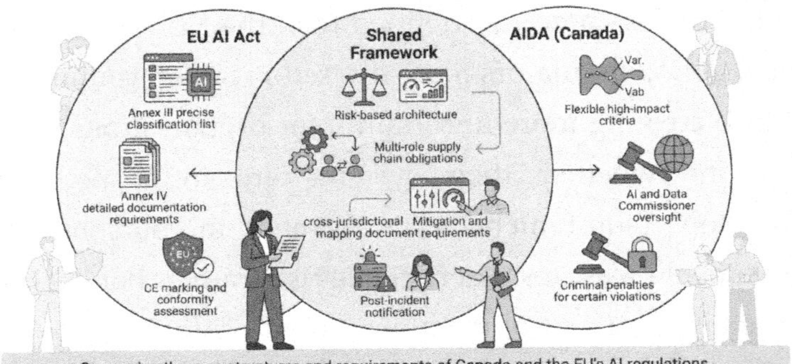

Comparing the core structures and requirements of Canada and the EU's AI regulations.

Compliance Planning for Organizations Operating Across the Canada-EU Axis

Organizations with significant operations in both Canada and the EU are best positioned to benefit from the conceptual overlap between AIDA and the EU AI Act. A compliance program built around EU AI Act requirements, including a comprehensive AI system inventory, documented risk classification, technical documentation for high-risk systems, and human oversight mechanisms, provides a strong foundation for AIDA compliance because it addresses most of the risk assessment and mitigation work that AIDA will require. The effort saved by building EU AI Act compliance first and adapting for AIDA second is substantially greater than building separate compliance programs for each jurisdiction.

The primary areas requiring jurisdiction-specific investment for Canada-EU compliance are documentation of the Canadian high-impact classification analysis (which requires applying AIDA's criteria rather than the EU Act's Annex III), the public disclosure requirement (which requires producing a description of each high-impact system's use in a format accessible to the public, not just to supervisory authorities), and the incident notification process (which requires engagement with the Canadian AI and Data Commissioner rather than EU national authorities). These are meaningful differences, but they do not require separate compliance infrastructure; they require adapting existing EU AI Act compliance processes and documentation to meet Canadian-specific requirements.

For organizations whose Canadian AI exposure is primarily in the financial services sector, the interaction between AIDA and the Office of the Superintendent of Financial Institutions (OSFI) guidance on model risk management adds a layer of complexity that is specific to the Canadian regulatory context. OSFI's expectations for model governance, model validation, and model documentation overlap substantially with those of AIDA and the EU AI Act. Still, OSFI also has specific expectations around model risk appetite, model tiering, and model inventory that reflect the Canadian financial sector's regulatory tradition. Financial institutions operating across the Canada-EU axis need to map their AI compliance program against all three frameworks simultaneously: EU AI Act, AIDA, and OSFI model risk guidance.

5.4 Brazil, China, and Emerging Regulatory Architectures

Beyond North America and Europe, enterprise AI compliance programs must contend with rapidly evolving regulatory frameworks in jurisdictions that collectively account for a significant share of global economic activity and AI deployment. Brazil and China represent two distinct models of AI regulatory development, each with different institutional architectures, different enforcement mechanisms, and different underlying theories of what AI governance is for. Understanding both is necessary for any global enterprise with operations in Latin America or Asia.

Brazil's AI Governance Framework and Its Sector-Specific Emphasis

Brazil's approach to AI governance has been developing through a combination of proposed national AI legislation and sector-specific regulatory action from existing authorities. The proposed Brazilian AI Bill has been debated in the National Congress since the early 2020s, with a focus on risk-based classification, the protection of fundamental rights, and accountability for AI systems used in consequential decisions affecting Brazilian residents. The legislative approach draws conceptual influence from the EU AI Act while reflecting Brazil's own constitutional framework, which places explicit protections for human dignity and non-discrimination at the center of fundamental rights analysis. When enacted, Brazilian AI legislation is expected to adopt the EU AI Act's high-risk category structure while adapting it to the Brazilian legal and institutional context.

In the interim, sector-specific regulatory action has been the primary mechanism for AI governance in Brazil. The National Data Protection Authority (ANPD), established under Brazil's Lei Geral de Proteção de Dados (LGPD, Brazil's data protection law modeled substantially on GDPR), has issued guidance on the use of automated decision-making in contexts involving personal data. LGPD grants data subjects the right to review decisions made solely on the basis of automated processing that significantly affects their interests, a right that the ANPD has been developing through regulatory guidance into something that parallels GDPR's Article 22 automated-decision provisions. Organizations that process personal data of Brazilian residents using AI systems that produce consequential automated decisions are subject to ANPD oversight under existing LGPD provisions, regardless of the status of AI-specific legislation.

For enterprise managers with Brazilian operations, the most important near-term compliance posture is to treat LGPD automated-decision requirements as current law and build compliance accordingly. This means documenting which AI systems produce or substantially inform automated decisions affecting Brazilian residents, ensuring that data subjects have access to meaningful human review of those decisions upon request, and maintaining records of the automated-decision process sufficient to explain the basis for decisions if challenged. These obligations apply now under the existing LGPD and provide a compliance foundation that will require relatively modest additional investment to satisfy more comprehensive AI-specific legislation when it is enacted.

Diagram 3.6 - Brazil AI Governance Structure

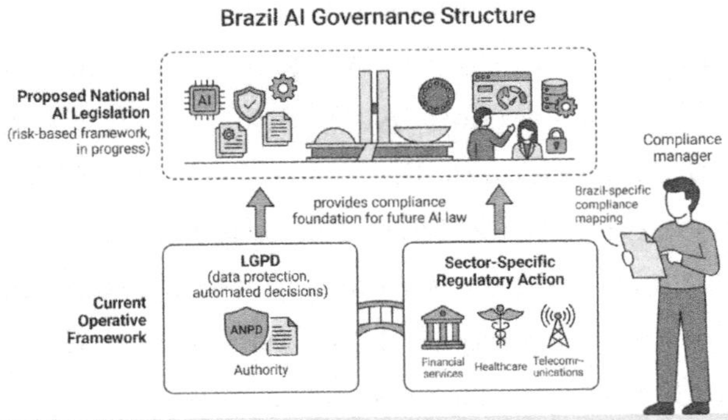

A structured overview of Brazil's current digital regulations and the emerging framework for Artificial Intelligence

China's Algorithmic Recommendation, Deep Synthesis, and Generative AI Regulations

China has taken a different regulatory path from Europe and North America, developing AI governance through a series of sector-specific and technology-specific regulations rather than through a single comprehensive framework. The three regulations most significant for enterprise compliance are the Provisions on the Management of Algorithmic Recommendations (effective March 2022), the Provisions on the Management of Deep Synthesis Internet Information Services (effective January 2023), and the Interim Measures for the Management of Generative Artificial Intelligence Services (effective August 2023). These three regulations together cover the major AI application categories that multinational enterprises are most likely to deploy in Chinese markets: recommendation systems, synthetic media, and generative AI.

The Algorithmic Recommendation Provisions apply to organizations that use algorithmic recommendation technology to provide information services to users within China. They require providers to disclose that algorithmic recommendations are being used, to provide users with options to view non-personalized content or to turn off personalized recommendations, and to prevent the use of algorithms to target users based on their consumption habits in ways that increase prices or engage in discriminatory pricing. Organizations operating e-commerce platforms, content delivery services, or customer communication platforms in China must audit their recommendation systems against these requirements and implement the mandated user control mechanisms.

The Generative AI Measures impose requirements that are particularly significant for enterprises using large language models in Chinese-market applications. Providers of generative AI services to the public in China must conduct security assessments before launching services, ensure that generated content complies with Chinese laws and socialist core values, implement measures to prevent the generation of false information, and require real-name registration for users. For multinational enterprises that use global generative AI platforms to power customer-facing applications in the Chinese market, these requirements pose a compliance challenge: the global platform's content policies may not meet Chinese content requirements, and the real-name registration requirement may conflict with the platform's global data governance practices. Separate Chinese-market AI configurations, potentially using China-approved AI providers, may be required to achieve compliance.

Diagram 3.7 - China's Layered AI Regulation Stack

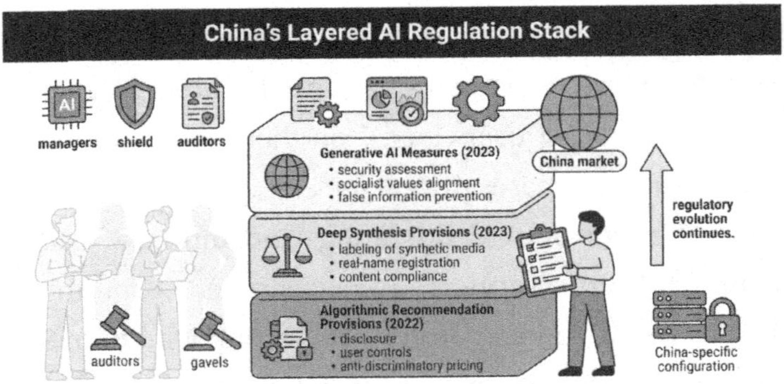

China enforces a layered regulatory framework to govern various applications of artificial intelligence.

What Divergent Regulatory Theories Mean for Global Product Design

The regulatory frameworks described above reflect different underlying theories of what AI governance is for. The EU AI Act is grounded in fundamental rights protection and market harmonization: it aims to prevent AI from violating individual rights while creating a consistent internal market in which products that comply with EU standards can circulate freely. Canada's AIDA is similarly rights-oriented, with an additional emphasis on accountability through a designated federal commissioner. US state laws are primarily focused on preventing discrimination in specific high-stakes decision-making contexts and on creating individual rights to explanation and correction.

China's framework reflects a different regulatory theory: it is primarily oriented toward information integrity, social stability, and alignment with national policy priorities. The requirement that generative AI systems align with socialist core values, the real-name registration requirements, and the security assessment process for new services all reflect a governance model in which the state's interest in controlling information flows is a primary compliance objective, alongside harm prevention. This divergence is not merely abstract: it has concrete implications for product design. An AI-powered content moderation system designed to comply with EU requirements (emphasizing transparency, human oversight, and non-discrimination) and a system designed to comply with Chinese requirements (emphasizing content alignment with national policy) may require fundamentally different architectures.

For product managers and architects, the divergent regulatory theories create a decision point that needs to be addressed at the product design stage: build a single global product with jurisdiction-specific configuration or build separate products for different regulatory environments. The single-global-product approach has obvious efficiency advantages but requires that the product's architecture be flexible enough to satisfy all applicable regulatory requirements simultaneously. The separate-product approach is more costly to maintain. Still, it may be necessary where regulatory requirements are genuinely incompatible at the architectural level rather than merely different in their specific content. Making this decision correctly requires regulatory, technical, and commercial expertise working in combination, and it needs to happen before significant development investment is made, not after the product is ready for launch.

Diagram 3.8 - Divergent Regulatory Theory Map

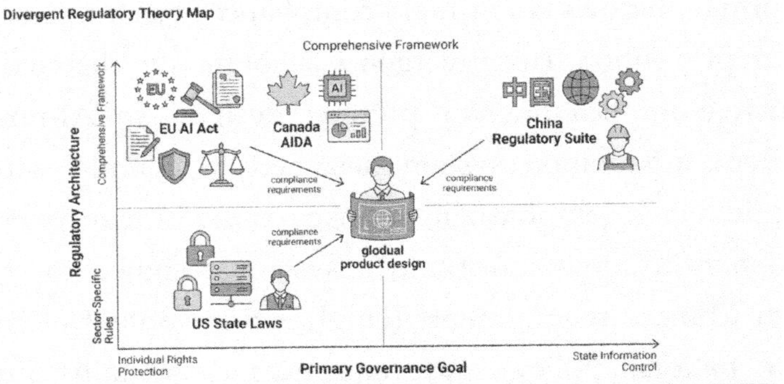

Mapping global approaches to AI governance and their impact on product design.

5.5 Managing Jurisdictional Complexity at Scale

The regulatory landscape described in this chapter is not a problem that can be solved once and for all. It is a dynamic environment that requires ongoing monitoring, periodic reassessment, and the organizational capacity to translate regulatory changes into compliance actions across a global enterprise. Organizations that manage this complexity effectively do so not by reacting faster than their peers to regulatory developments but by building structural capabilities that make monitoring, analysis, and adaptation systematic rather than reactive.

Regulatory Mapping: Building a Living Inventory of Applicable Laws

A regulatory map is the foundational instrument for managing multi-jurisdictional AI compliance. It is a maintained record of every AI-specific law, regulation, and binding guidance that applies to the organization's AI systems, organized by jurisdiction, system type, and effective date. A well-constructed regulatory map answers several questions compliance managers routinely need: Which of our AI systems are subject to which regulatory frameworks? What are the specific obligations in each framework that apply to each system? When do those obligations take effect? What is the current status of our compliance with each obligation?

Building the regulatory map requires an initial investment in legal research across all jurisdictions where the organization has operational AI deployments or users. This research cannot be completed in-house for most organizations: it requires jurisdiction-specific legal expertise in each relevant market, because AI regulatory interpretation requires understanding not just the text of the law but also the regulatory culture, enforcement history, and interpretive guidance of each jurisdiction. External counsel with AI regulatory expertise in each major jurisdiction is a standard component of the regulatory mapping exercise for global enterprises, and the output of that counsel's work needs to be integrated into a single, maintained instrument rather than distributed across separate jurisdiction-specific memos.

The 'living' aspect of the regulatory map is as important as its initial construction. The AI regulatory landscape is changing at a pace that makes static regulatory maps obsolete within months of their production. An organization that completed a regulatory map in early 2024 and has not updated it has missed the enactment of Colorado's AI Act, the progression of Canada's AIDA through legislative review, updates to China's generative AI guidance, and the first round of EU AI Act implementation guidance from the AI Office and national competent authorities. Maintaining the map as a living document requires a designated owner with responsibility for monitoring regulatory developments across all relevant jurisdictions, a process for triaging new developments by urgency and materiality, and a regular review cycle that ensures the map reflects the current state of the regulatory landscape.

Diagram 3.9 - Living Regulatory Map Structure

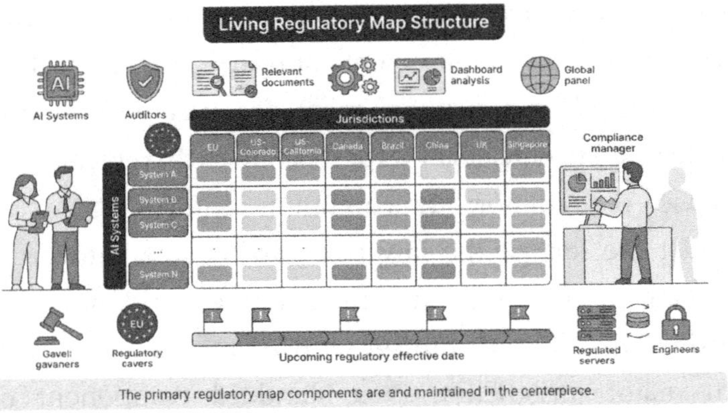

The Case for a Unified Compliance Architecture Versus Jurisdiction-Specific Approaches

The choice between unified compliance architecture and jurisdiction-specific compliance programs is among the most consequential design decisions in global AI compliance management. A unified architecture identifies the highest common denominator requirements across all applicable frameworks. It builds compliance processes and documentation that satisfy all of them simultaneously, with jurisdiction-specific supplements for requirements that are unique to individual frameworks. A jurisdiction-specific approach builds separate compliance programs for each regulatory environment, optimized to each framework's specific requirements without attempting to rationalize across them.

The case for a unified architecture rests on efficiency and consistency. Most of the compliance work required by the major global AI frameworks is conceptually similar: inventory AI systems, classify them by risk, assess them for compliance with applicable obligations, document the compliance evidence, implement oversight mechanisms, and monitor ongoing performance. A unified architecture that performs this work once and applies the results across multiple frameworks is substantially less resource-intensive than a series of parallel processes that each perform similar work in different formats. It also produces more consistent governance outcomes: an AI system that has been assessed once under a rigorous, unified framework is less likely to have material compliance gaps than a system that has been assessed multiple times under different frameworks by different teams, with different documentation standards and interpretive benchmarks.

The case for jurisdiction-specific approaches rests on precision and risk management. Unified architectures work best when the underlying frameworks share enough structural similarity that a single compliance process can address all of them without requiring significant per-framework adaptation. When frameworks differ structurally, as China's information-governance-oriented framework does from the EU's rights-protection framework, a unified architecture may sacrifice precision across individual jurisdictions. For organizations where regulatory enforcement risk in a specific jurisdiction is high (due to the size of the market, the sensitivity of the systems deployed, or the enforcement activity of the local authority), the cost of imprecision may outweigh the efficiency gains of a unified approach to that jurisdiction's requirements. The practical answer for most global enterprises is a unified core architecture with jurisdiction-specific supplements, using the unified core for the compliance work that all frameworks share and building jurisdiction-specific processes for requirements that cannot be satisfied through the unified approach.

Diagram 3.10 - Unified vs. Jurisdiction-Specific Architecture

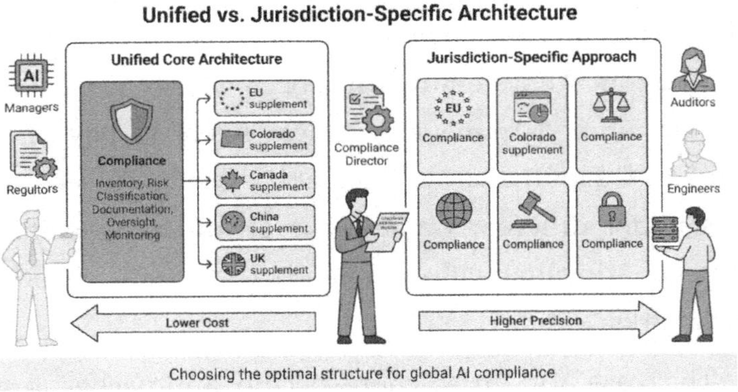

Monitoring Regulatory Development and Anticipating Convergence or Divergence

Anticipating where the regulatory landscape is heading is as important as understanding where it is now. Organizations that monitor regulatory development strategically, rather than reactively, can make compliance investments that are durable across regulatory evolution rather than investments that need to be replaced when the next wave of regulatory change arrives. The key monitoring inputs are legislative pipelines (bills introduced, committee activity, public consultations), regulatory guidance publications (interpretive notes, enforcement policies, supervisory Q&A documents), enforcement actions (which reveal enforcement priorities and interpretive standards), and international policy forums (where regulatory convergence or divergence is negotiated at the intergovernmental level).

Convergence signals indicate that regulatory frameworks across jurisdictions are moving toward a common standard. Convergence tends to occur when regulatory bodies participate in common international forums, when a framework with high extraterritorial reach (like the EU AI Act) creates market pressure toward its standards, or when a major enforcement action in one jurisdiction creates political and institutional pressure in others to adopt similar requirements. Divergence signals indicate that regulatory frameworks are moving in different directions, either because of different underlying regulatory theories (as between EU and Chinese frameworks) or because of jurisdiction-specific political dynamics that override convergence pressures.

For a global enterprise, Singapore's approach to AI governance provides a useful case study in monitoring regulatory evolution. Singapore introduced its Model AI Governance Framework in 2019 as a voluntary, principles-based guide, similar to the NIST AI RMF in the United States. It developed this into the AI Verify testing toolkit, which allows organizations to test their AI systems against specific governance criteria. Singapore subsequently joined the development of a multilateral AI governance framework through the Global Partnership on AI and the ASEAN AI Governance Framework. The trajectory of Singapore's approach, from voluntary principles to testing toolkit to international governance engagement, illustrates the direction that jurisdictions at earlier stages of regulatory development typically follow. Organizations monitoring the Singapore trajectory are better positioned to anticipate when it will move toward binding law, and what that binding law is likely to require, based on what the voluntary framework has established as the normative standard.

5.6 Manager's Checklist

The following items represent the actions required to build a multi-jurisdictional AI compliance program that can absorb regulatory change without requiring full reconstruction each time a new framework enters into force, or an existing framework is updated. Each item is verifiable through a document, an organizational structure, or a defined process.

- Confirm that your organization has completed a cross-jurisdictional regulatory mapping exercise that identifies every AI-specific law and binding guidance applicable to your AI system portfolio, organized by system, jurisdiction, and effective date.
- Verify that your portfolio has been assessed for Colorado AI Act scope (consequential decisions affecting Colorado residents), and that algorithmic impact assessments have been initiated for all systems determined to be in scope, with completion targeted before the February 2026 effective date.
- Confirm that your organization has identified all AI systems used in employment decisions affecting Illinois residents and assessed them against the Artificial Intelligence Video Interview Act and

applicable BIPA provisions, with consent and notification mechanisms documented.

- Verify that any AI systems used in California-resident-facing consequential decisions have been assessed against CPRA automated-decision-making rights. That explanation and opt-out mechanisms are operational for in-scope systems.
- Confirm that your organization has assessed its AIDA exposure for Canadian operations, identified which AI systems are likely to qualify as high-impact systems, and established a monitoring process for AIDA enactment and implementation timelines.
- Verify that AI systems deployed in Brazilian markets that produce or substantially inform automated decisions affecting Brazilian residents have been assessed against LGPD automated-decision requirements, with ANPD-compliant human review mechanisms documented.
- Confirm that AI systems deployed in Chinese markets (including recommendation systems, synthetic media tools, and generative AI applications) have been assessed against the applicable Chinese AI regulations, and that required security assessments, real-name registration, and content compliance measures are operational.
- Verify that a designated regulatory monitoring function exists with defined responsibility for tracking AI regulatory developments across all jurisdictions in scope, a defined triage process for assessing materiality

of new developments, and a defined cadence for updating the organizational regulatory map.

- Confirm that your organization has made the unified-versus-jurisdiction-specific architecture decision explicitly, with a documented rationale, and that the chosen architecture has been implemented consistently across all AI systems in scope for multi-jurisdictional compliance obligations.

5.7 Takeaway

The global AI regulatory landscape is best understood not as a collection of separate local problems but as a single multi-dimensional compliance challenge that requires a single, coherent compliance architecture capable of addressing multiple frameworks simultaneously. Organizations that build their compliance programs around a single framework and treat others as extensions or add-ons consistently find that retrofit costs are higher than the cost of building for multi-jurisdictional compliance from the start. The unified architecture approach, with jurisdiction-specific supplements for requirements that cannot be addressed through the common core, is the structural design that produces the best balance of efficiency, precision, and durability over time.

The most important single capacity for managing multi-jurisdictional AI compliance is not legal expertise in any single framework: it is the organizational capability to continuously monitor regulatory evolution, assess the implications of changes for the current compliance program, and translate those implications into operational decisions before effective dates create enforcement exposure. That capability is built on a living regulatory map, a designated monitoring function, and an organizational culture in which regulatory intelligence is treated as a strategic input to AI product and deployment decisions rather than a post-hoc constraint. Managers who build that culture are building not just compliance capacity for the current regulatory environment but resilience for the one that follows it.

6 Compliance as Organizational Discipline

A vice president of engineering at a mid-sized European insurance technology company walked into a board meeting with a one-hundred-and-twelve-page AI policy document and left two hours later with an action plan to rebuild her compliance program from the ground up. The board had asked one question: 'Can you show us the last time a model decision was challenged, who reviewed it, and what the outcome was?' She could not answer. The policy document existed. The governance committee met twice. The chief data officer had signed a framework. But the company had no decision log, no escalation record, no audit trail showing that any of the policies had actually been applied to any of the nine AI systems currently deployed in claims processing and underwriting. The organization had compliance documentation without compliance capability, and the distinction was about to become legally significant.

That gap, between having a policy and having an operational discipline, is the defining challenge of AI compliance in the current regulatory environment. The EU AI Act does not reward organizations for producing impressive governance frameworks. It rewards organizations that can demonstrate, with artifacts and records, that their AI systems meet substantive requirements at every stage of the lifecycle. Building that demonstration capability requires coordinated change across people, processes, and technology, and it requires treating compliance not as a project with an end date but as a standing feature of how the organization runs.

6.1 Why It Matters

The EU AI Act is not a one-time certification exercise. It imposes ongoing obligations: continuous monitoring, updated technical documentation, post-market surveillance, serious incident reporting, and re-assessment whenever a system undergoes a substantial modification. Organizations that treat compliance as a project, complete it once, and move on will find themselves out of conformity within months of their initial certification. The Act is designed to create sustained organizational capability, not a snapshot of good behavior.

For managers, this means that the resources required for AI compliance are not a temporary budget spike but a permanent operational cost that must be built into team structures, hiring plans, tool budgets, and process designs. It also means that the return on building genuine compliance capability is compounding: organizations that internalize compliance discipline accelerate their own AI deployment cycles because they stop debating whether a new system needs governance attention and instead have clear workflows for processing it. Every AI initiative that enters a functioning governance pipeline moves faster than it would in an organization where compliance is improvised.

The three-pillar framing of people, processes, and technology is not a management consulting abstraction. It reflects a structural reality about how organizations embed any new operational requirement. New regulations fail to take hold when organizations address only one or two pillars. A team of compliance professionals without supporting processes produces inconsistent outcomes. Documented processes without technical tooling remain aspirational. Technical tooling without trained personnel generates data that no one acts on. The discipline exists only when all three pillars are in place and connected.

6.2 From Policy to Practice: The Compliance Implementation Gap

Most organizations entering their first serious AI compliance cycle discover the same uncomfortable truth: they have produced policies that describe desired behavior in abstract terms but have not connected those policies to the specific decisions, workflows, and artifacts that constitute actual AI development and deployment. The gap between policy and practice is not a failure of intent; it is a failure of implementation design.

Why Documentation-Heavy Programs Fail at the Operational Level

A compliance program built around documents rather than behaviors fails for a predictable reason: documents describe what should happen, not what does. Still, they cannot enforce it, detect when it does not happen, or record what actually did happen. The EU AI Act's requirements are behavioral requirements. The Act asks whether a risk management system was applied to a specific model, whether training data was documented with specific attributes, and whether human oversight was meaningful or perfunctory. None of those questions can be answered by pointing to a policy statement.

Organizations that have navigated mature compliance regimes in adjacent domains, such as financial services, prudential regulation, or medical device quality management, understand this distinction intuitively. They build compliance into workflows by making compliant behavior the path of least resistance. An AI system cannot be deployed without a completed risk classification record. A model cannot enter production without a sign-off from the AI governance committee. An AI vendor product cannot be approved without a completed supplier assessment. When non-compliance requires affirmative action to bypass a workflow gate, it becomes rare. When compliance requires extra effort beyond normal work, it remains inconsistent.

The operational failure mode most relevant to the EU AI Act is what internal audit teams call 'policy shelf risk': the policy exists, it was approved, someone was assigned responsibility for it, but no one changed any actual workflow to implement it. Six months later, auditors find that every team assumed some other team was handling it. The corrective action is not to update the policy. It is to map each policy requirement to a specific workflow touchpoint, a specific role that is accountable for executing it, and a specific artifact that proves it was done.

Managers leading compliance implementation should begin by conducting a workflow-level gap analysis rather than a document-level one. The question to ask for each AI Act requirement is not 'do we have a policy that covers this?' but 'in which workflow does this requirement get met, who does it, and what is the evidence?' If those three answers cannot be filled in, the requirement has not been implemented, regardless of what the policy document says.

A useful implementation diagnostic is to trace a single AI system through its entire lifecycle and ask at each stage what compliance evidence was generated and where it is stored. For most organizations, this exercise surfaces the same gaps: risk classification was informal and undocumented, training data provenance was tracked in a spreadsheet no one maintains, and human oversight was described in an architecture document but not operationalized in any deployment runbook. The trace exercise produces an inventory of missing artifacts that becomes the implementation backlog.

The diagram below illustrates the gap between policy-layer compliance and workflow-layer compliance, showing how behavioral checkpoints and artifact requirements sit at the level of actual development decisions rather than policy statements.

One additional failure mode deserves attention: compliance responsibility concentrated in a single team. When AI compliance is treated as a legal or risk function that reviews systems after they are built, two things happen consistently. First, the legal and risk team is always overwhelmed, becoming a bottleneck. Second, development teams optimize for getting past the review rather than for building compliant systems, which produces systems that satisfy documentation requirements without satisfying substantive ones. Distributed accountability, where each function owns the compliance aspects of its own work, is the only model that scales.

Diagram 4.1 - Policy Layer vs. Workflow Layer Compliance

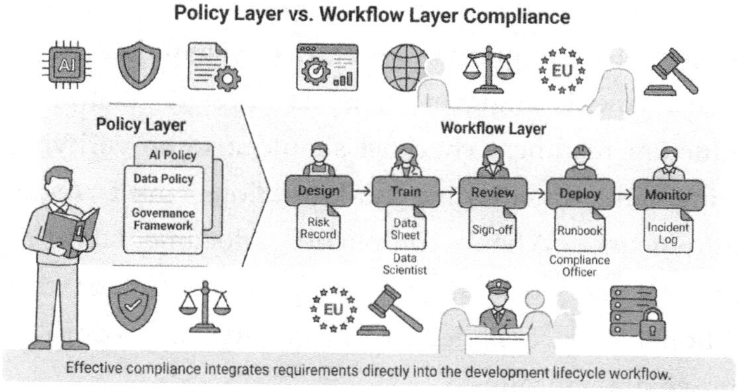

Embedding Compliance Requirements into Development and Procurement Workflows

Embedding compliance into development workflows means identifying every decision point in the AI lifecycle where a compliance requirement applies and designing the workflow so that the requirement is met as part of normal work, not as an add-on. For development teams, this typically means integrating compliance gates into the existing software development lifecycle: adding a risk classification step to the intake process for new AI initiatives, adding a documentation requirement to the model card or system card generated during development, adding a human oversight verification step to deployment approval, and adding monitoring configuration to the production handoff checklist.

The key design principle is that compliance artifacts should be generated by the people who are already doing the underlying work, using the tools they already use, at the moment the underlying work is being done. A data scientist who is already documenting their training dataset should also be documenting the data governance attributes required by the Act. A model developer who is already running evaluation benchmarks should also be running the accuracy and robustness tests required for high-risk system documentation. A deployment engineer who is already completing a production readiness checklist should also be verifying that logging and monitoring configurations meet regulatory requirements. When compliance documentation is a byproduct of work already being done, rather than an additional task assigned later, quality and completeness improve dramatically.

Procurement workflows require a different but equally systematic approach. Every AI component, model, or system procured from an external vendor introduces compliance obligations that must be negotiated and verified before the purchase is made. Supplier due diligence for AI procurement should include verification that the vendor can provide technical documentation for the AI components being supplied, that the vendor's quality management system has been assessed, that contractual provisions assign responsibility for maintaining documentation and notifying the deployer of substantial modifications, and that the vendor's conformity assessment status is current. Many organizations discover during their first Act-era procurement cycle that their existing vendor contracts contain no AI-specific provisions at all and must be renegotiated or supplemented.

Diagram 4.2 - AI Development Lifecycle Compliance Gates

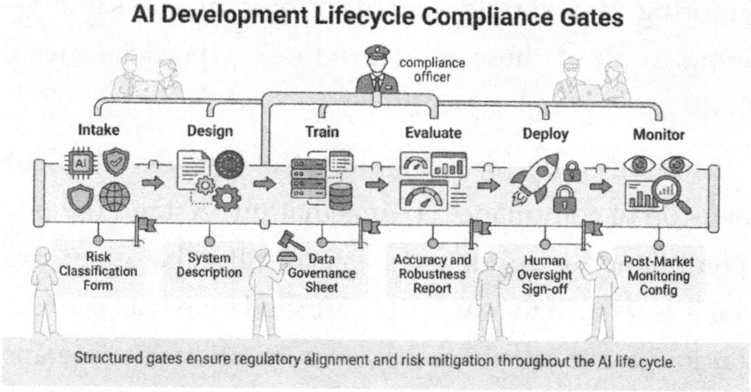

The Role of Training and Awareness in Building a Compliance Culture

Technical workflows and governance structures will not function if the people executing them do not understand what they are doing or why it matters. Training and awareness programs for AI compliance have a different character than conventional compliance training because the subject matter is still evolving, the requirements are technically complex, and the audience spans a wide range of roles with very different relationships to AI systems.

Role-differentiated training is the correct starting model. A business owner approving a new AI initiative needs to understand the risk classification criteria to identify systems that require additional governance attention. A data scientist preparing training data needs to understand the Act's data governance requirements with sufficient detail to generate compliant documentation. A procurement manager negotiating an AI vendor contract needs to understand what documentation rights and responsibilities must be captured in the agreement. A legal or risk professional reviewing a deployed system needs to understand what the post-market monitoring plan should contain. Providing the same generic training to all of these roles produces superficial awareness without operational capability.

Training programs should also address the cultural dimension of compliance. Teams that understand the purpose of compliance requirements, specifically that they exist to prevent real harm to real people affected by AI decisions, make better judgment calls in ambiguous situations than teams that view compliance as a bureaucratic imposition. This is particularly important for the EU AI Act because many of the Act's requirements involve qualitative judgment rather than quantitative threshold-checking. What constitutes 'meaningful' human oversight? What makes a training dataset 'relevant, representative, and free of errors'? These questions cannot be answered by reference to a checklist alone; they require teams to exercise informed judgment, and that judgment improves with a genuine understanding of the regulatory purpose.

Awareness mechanisms should complement formal training: regular compliance updates during team meetings, brief reminders in project management tools when AI-relevant milestones are reached, internal case studies based on the organization's compliance experience, and easy access to the internal policy stack through knowledge management systems. The goal is to make compliance knowledge ambient rather than episodic, available at the moment of need rather than only recalled from a training session attended six months ago.

6.3 People: Roles, Responsibilities, and Accountability Structures

The EU AI Act does not mandate specific organizational structures or job titles. What it does do is create a set of obligations that must be owned by specific people with specific authority to act. Organizations that distribute those obligations across existing roles without adjusting accountability structures find that obligations fall through the cracks precisely because no one's performance is evaluated on whether they are met.

The Compliance Roles the EU AI Act Implicitly Requires

Four functional roles are essential for compliance with the EU AI Act, regardless of how an organization assigns them across its org chart. The first is the AI governance authority: a person or committee with the standing to approve or block AI system deployments based on compliance criteria. Without this role, compliance review has no teeth. The second is the technical documentation owner: a person responsible for the completeness and accuracy of the technical file for each high-risk AI system. This person may be a senior AI engineer, a quality engineer, or a dedicated compliance engineer, but the accountability must be explicit. The third is the risk assessment lead: a person with the expertise to conduct and document the risk management activities required by the Act, typically someone with a background in risk management combined with AI technical literacy. The fourth is the regulatory interface: a person responsible for interactions with market surveillance authorities, for managing serious incident notifications, and for maintaining the organization's registration in the EU database for high-risk AI systems.

In large organizations, each of these roles may be filled by dedicated staff. In mid-sized and smaller organizations, roles will be combined. A chief AI officer may hold both the governance authority role and the regulatory interface role. A lead AI engineer may serve as the technical documentation owner for all systems in their domain. The critical requirement is not headcount but clarity of accountability. Each role must be assigned, the assignment must be documented, and the assigned person must have both the authority and the resources to fulfill the role.

The EU AI Act's provider-deployer distinction creates an additional accountability dimension for organizations that deploy AI systems built by external providers. In that situation, the deployer assumes specific obligations, including conducting a fundamental rights impact assessment in certain circumstances, implementing the provider's human oversight instructions, and maintaining logs of system operations. Those deployer obligations must be assigned to specific internal roles with access to the system in production. Many organizations discover that their internal accountability structures were designed to build AI rather than deploy it, and must be extended to cover the ongoing obligations that arise from operating AI systems developed elsewhere.

The Chief AI Officer (CAIO) role has emerged as the most common organizational response to the need for a senior accountability anchor for AI governance. A functioning CAIO role, whether it is a dedicated executive position or an additional portfolio for an existing executive, requires genuine authority over AI system approvals, a direct reporting line to the board or the CEO, and a budget for compliance infrastructure. A CAIO who can observe compliance failures but cannot block deployments has no operational function. Organizations should be clear-eyed about whether their CAIO role has real authority or is merely a communications designation.

Diagram 4.3 - Core AI Compliance Role Map

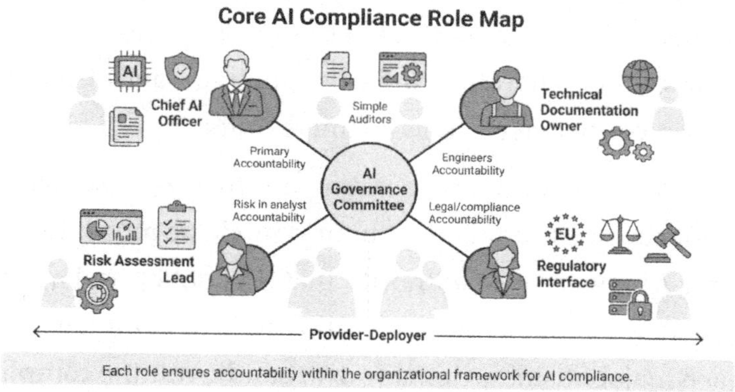

Distributing AI Compliance Responsibility Across Functions

Effective AI compliance is a cross-functional discipline. The roles described above require ongoing coordination with engineering, data science, legal, risk, procurement, business operations, and product management. The question of how to structure that coordination is one of the most consequential organizational design decisions an enterprise faces in the current compliance environment.

The two dominant models are a centralized compliance team model and a federated compliance model. In the centralized model, a dedicated AI governance team holds primary responsibility for compliance across all AI systems and functions as the reviewer and approver for all compliance documentation. In the federated model, compliance responsibility is distributed among function-level compliance leads embedded in engineering and business teams, with a central governance team providing standards, tools, and oversight. The centralized model produces consistency but creates bottlenecks and can create an adversarial dynamic between development teams and the compliance function. The federated model scales better but requires substantial investment in shared standards, tooling, and training to maintain consistency across business units.

For most enterprises operating multiple AI systems across multiple business domains, a hybrid model is most practical: a central AI governance authority that sets policy and approves high-risk and novel systems, with federated compliance leads in each major business domain who handle routine compliance activities for systems in their domain and escalate to the central authority when they encounter novel situations. This structure mirrors the approach that mature regulated industries, such as pharmaceutical manufacturing and financial services, have developed over decades of multi-jurisdictional compliance management.

The MLOps function, responsible for the operational deployment and monitoring of AI models, sits at a uniquely important intersection in this structure. MLOps engineers control the infrastructure that generates compliance evidence: logging systems that record model decisions, monitoring dashboards that surface performance drift, and deployment pipelines that enforce compliance gates. Compliance programs that treat MLOps as a purely technical function and do not integrate compliance requirements into MLOps workflows will find that their compliance infrastructure is technically sophisticated but compliance-blind. MLOps teams need clear guidance on the compliance requirements for each system they operate and on the tools to generate and preserve evidence demonstrating compliance.

Diagram 4.4 - Federated Compliance Accountability Model

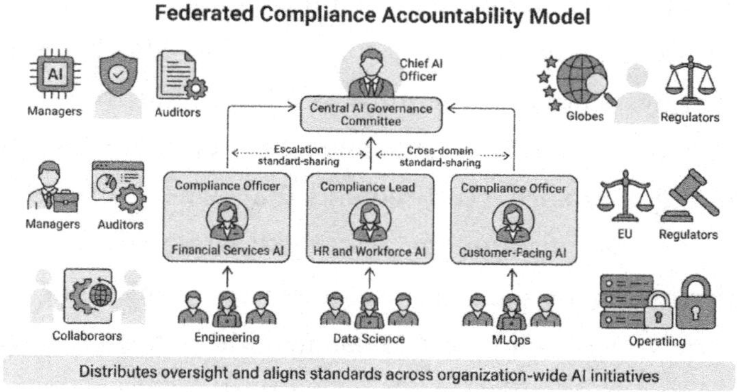

Accountability Mechanisms: Who Is Responsible When Something Goes Wrong

Clear accountability at the role level is necessary but not sufficient. Organizations also need escalation protocols that specify what happens when a compliance issue is identified, who has authority to make which decisions, and what record is generated by the resolution process. Without explicit escalation protocols, organizations find that compliance issues are either informally dismissed by people without authority or unnecessarily escalated to senior leadership for decisions that could be handled at a lower level.

A well-designed escalation protocol distinguishes between three levels of compliance issues. The first level covers routine compliance gaps that can be resolved within normal development or operations workflows: an incomplete documentation field that can be completed, a monitoring alert that was not configured and can be added, and a missing training record that can be obtained. These issues are tracked in the compliance management system and resolved by the responsible team without formal escalation. The second level covers material compliance concerns that affect a system's conformity status or require a change to the system's deployment scope: a risk management step that was skipped, a data governance requirement that was not met, or a substantial modification that may require re-assessment. These require review by the technical documentation owner and the risk assessment lead, and must include a formal deviation record and a remediation plan. The third level covers serious compliance failures that could affect regulatory standing or require supervisory notification: a serious incident involving the AI system, a finding that a deployed system does not meet the requirements of its conformity assessment, or a decision to withdraw a system from the market. These require immediate involvement of the regulatory interface role and the CAIO, with documented escalation to board level if warranted.

The escalation protocol should be documented in a runbook that specifies who initiates the escalation, who is notified, the decision timeline, and the artifacts generated at each stage. The runbook should be tested in tabletop exercises at least annually so that the people involved know their roles before a real incident occurs. Organizations that have never tested their escalation protocol discover its gaps in the worst possible moment.

Diagram 4.5 - Three-Level Compliance Escalation Protocol

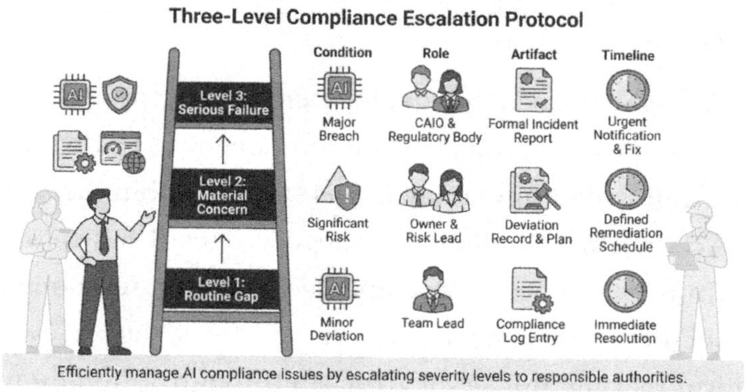

6.4 Process: Governance Workflows for the AI Lifecycle

Governance processes give accountability structures their operational form. A well-designed governance process converts a compliance requirement into a sequence of steps, a list of people who must be involved, a set of artifacts that must be produced, and a record of completion. Without documented governance processes, compliance depends on individual judgment, varies across teams, and cannot be demonstrated to external auditors.

Pre-Deployment Review Gates and AI System Approval

The most important governance process in the AI compliance lifecycle is the pre-deployment review gate: the formal checkpoint at which an AI system's compliance status is assessed before it is placed in service. Pre-deployment review serves two functions. It prevents non-compliant systems from being deployed, which avoids regulatory exposure and harm to affected individuals. It also creates a documented record that compliance was assessed at the point of deployment, which is essential evidence in any subsequent audit or investigation.

A well-designed pre-deployment review process for EU AI Act purposes includes several components. The first is a risk classification determination: is this system within the scope of the Act, and if so, what risk category does it fall into? This determination must be documented with the reasoning, not merely the conclusion. For high-risk systems, the review must confirm that the full set of Article 9 through 15 requirements has been addressed. The second component is a technical documentation review: Is the technical file complete, current, and consistent with the system as built and tested? The third is a human-oversight verification: has the human-oversight mechanism described in the technical file been implemented in the deployed system and tested? The fourth is a post-market monitoring readiness check: is the monitoring configuration in place, are the logging requirements met, and are the performance thresholds and alert conditions documented?

The output of the pre-deployment review should be a signed approval record that captures the date, the reviewer identities, the risk classification, the key compliance findings, any open items with remediation timelines, and the approval decision. This record becomes part of the system's compliance file and is the primary artifact that demonstrates the organization performed pre-deployment due diligence.

Organizations with large AI portfolios face the practical challenge of applying consistent pre-deployment review standards across systems of very different types and risk levels. A tiered review protocol addresses this: minimal-risk systems that are clearly outside the scope of the Act receive a brief risk classification check and are passed through without detailed review; limited-risk systems receive a standard review focused on compliance with transparency requirements; high-risk systems receive the full review with external sign-off from the AI governance committee. The risk classification step at intake filters systems to the appropriate review track, preventing both under-review of genuinely risky systems and over-review of routine low-risk ones.

The pre-deployment review process should be supported by a review checklist that is specific enough to be applied independently by different reviewers and to produce consistent outcomes. Generic checklists that ask questions like 'have all compliance requirements been met?' produce compliance theater rather than compliance assurance. Specific checklists that ask 'is Section 3.2 of the technical file completed with the information specified in Annex IV, item 2b?' produce verifiable evidence.

Diagram 4.6 - Tiered Pre-Deployment Review Process

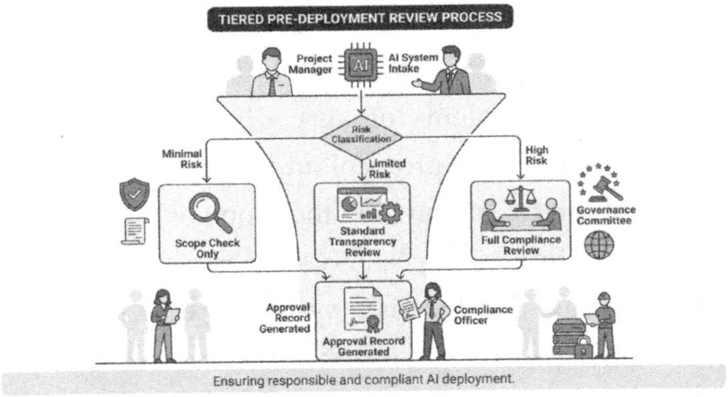

Incident Detection, Escalation, and Remediation Protocols

Post-deployment incident management is a compliance obligation, not merely a quality management one. The EU AI Act requires providers of high-risk AI systems to report serious incidents to market surveillance authorities within specified timelines: fifteen days for incidents involving death or serious harm, three days for the most severe cases. Meeting those timelines requires incident-detection capability, a clear definition of what constitutes a reportable incident, a documented escalation process, and a preparation protocol for the notification content.

The EU AI Act defines a serious incident as a malfunction or performance deficiency that directly or indirectly leads to the death or serious injury of a person, a serious breach of obligations under Union law intended to protect fundamental rights, or serious property damage. Organizations need to translate that definition into operational terms specific to their AI systems. For an AI system used in credit risk assessment, a serious incident could be a systematic error that affects a protected class. For an AI system used to support medical diagnosis, it might be a pattern of incorrect recommendations in a specific clinical context. The operational definition should be specific enough to be applied consistently by the people who monitor the system's performance.

Incident detection requires a monitoring infrastructure capable of surfacing anomalous system behavior in time to meet notification timelines. Many organizations discover that their monitoring systems are designed to detect technical failures, such as system outages and latency spikes, but are not designed to detect compliance-relevant performance issues, such as unexpected demographic disparities in model outputs or systematic degradation in specific input domains. Compliance-aware monitoring configuration is a technical requirement that must be specified by the risk and compliance function and implemented by the MLOps or operations team.

The remediation protocol should specify what happens after an incident is detected: immediate containment steps (which may include suspending the system), root cause analysis, documentation of the incident and its effects, notification to supervisory authorities if required, and the corrective action plan. Each step should have a responsible role and a timeline. The protocol should be exercised in a tabletop simulation before a real incident occurs, and the simulation findings should be used to identify gaps in detection capability, escalation speed, or documentation quality.

Diagram 4.7 - Incident Detection and Reporting Workflow

Incident Detection and Reporting Workflow

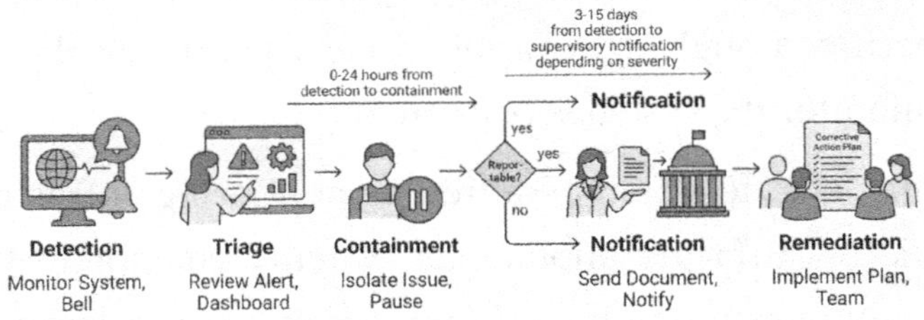

Streamlined processes for rapid response and regulatory compliance.

Supplier and Vendor Due Diligence for AI Procurement

Organizations that procure AI systems or AI components from external suppliers do not escape the Act's compliance obligations by outsourcing development. Deployers of high-risk AI systems take on specific obligations under the Act, and the quality of the technical documentation they receive from providers directly affects their ability to meet those obligations. Supplier due diligence is therefore not simply a procurement risk management practice; it is a compliance requirement embedded in the deployer's role.

A robust AI supplier due diligence process includes several elements. Before procurement, the organization should request the provider's conformity assessment documentation for the AI components under consideration, verify that the provider has a functioning quality management system for the AI product, confirm that the provider maintains and updates technical documentation, and assess the contractual terms governing notification of substantial modifications. During contract negotiation, the organization should ensure that the agreement contains provisions requiring the provider to notify the deployer of any substantial modifications, to maintain the technical documentation for the required period (ten years for many high-risk categories), to cooperate with supervisory authority requests for information, and to provide the deployer with the information needed to discharge its own obligations as a deployer.

After procurement, the supplier due diligence process does not end. The organization should periodically monitor the provider's compliance status, verify that technical documentation updates are received upon modifications, and assess whether any changes to the provider's product affect the organization's conformity status. For AI systems that are deeply embedded in critical operations, a supplier transition plan should be maintained if a provider's compliance status deteriorates.

Many organizations are discovering that the AI supplier landscape has not yet adapted fully to the Act's requirements. Providers of general-purpose AI components may not have the documentation or quality management infrastructure that deployers need, and deployers may need to negotiate documentation rights they did not previously have. Organizations that are early movers in requiring AI-specific contractual protections from suppliers will be better positioned as the market matures and supplier compliance becomes a competitive differentiator.

Diagram 4.8 - AI Supplier Due Diligence Process

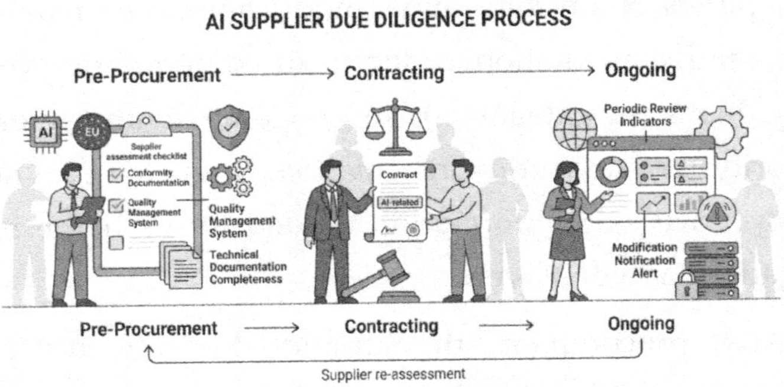

A systematic approach to ensuring AI vendor compliance and risk mitigation.

6.5 Technology: The Infrastructure of Compliance

Compliance governance at scale cannot run on spreadsheets and email. Organizations managing more than a handful of AI systems need purpose-built or adapted tooling to maintain model inventories, generate and store compliance documentation, monitor the performance of deployed systems, and produce audit-ready evidence. The technology infrastructure of compliance is not a luxury; it is the mechanism by which the people and processes described in the preceding sections can function consistently and efficiently.

Model Registries, Documentation Systems, and Audit Trail Requirements

A model registry is the foundational tool of AI compliance governance. At its core, a model registry is a structured inventory of every AI system the organization operates, maintains, or has placed in service, with sufficient information about each system to support compliance management. For EU AI Act purposes, the registry should capture the system name and description, the risk classification determination and its rationale, the conformity assessment status, the location of the technical documentation file, the owner and responsible team, the current deployment status and environment, the version history, the post-market monitoring configuration, and the date of the last compliance review.

A registry that contains only basic metadata about models is a model catalog, not a compliance registry. The distinction matters because compliance management requires the registry to be actionable: it should surface which systems are due for re-assessment, which systems have open compliance findings, which systems are approaching the modification threshold that would require re-classification, and which systems are subject to upcoming regulatory deadlines. A good compliance registry is a live management tool, not a static inventory.

The technical documentation system is a companion to the model registry. While the registry tracks the status and metadata of each system, the technical documentation system stores the actual compliance artifacts: the full technical file per Annex IV requirements, the risk management records, the testing and evaluation reports, the data governance documentation, the instructions for use, and the conformity assessment records. Access controls on the documentation system should preserve the integrity of the record, ensuring that documentation cannot be modified after the fact without an audit trail.

Decision logs and deviation records are compliance artifacts that many organizations overlook when designing their documentation infrastructure. A decision log captures the substantive governance decisions made about each AI system: why a particular risk classification was chosen, why a specific human oversight design was selected, and why a potential risk was accepted rather than mitigated. A deviation record documents cases where the standard compliance process was not followed and the approved justification for the deviation. Both types of records are essential for demonstrating that the organization's compliance program was exercised with genuine judgment, not merely formal compliance with a checklist.

Audit trail requirements under the EU AI Act vary by system type and risk level. For high-risk AI systems, the Act requires that logs be kept that allow for post-hoc verification of system operation, including the inputs and outputs used to make decisions affecting individuals. Log retention requirements extend up to six months by default, and longer where specific sector requirements apply. Organizations should ensure that their logging infrastructure can satisfy these requirements and that logs are stored in a format and location that allows retrieval within the timelines that supervisory authority investigations may require.

Diagram 4.9 - Compliance Documentation Architecture

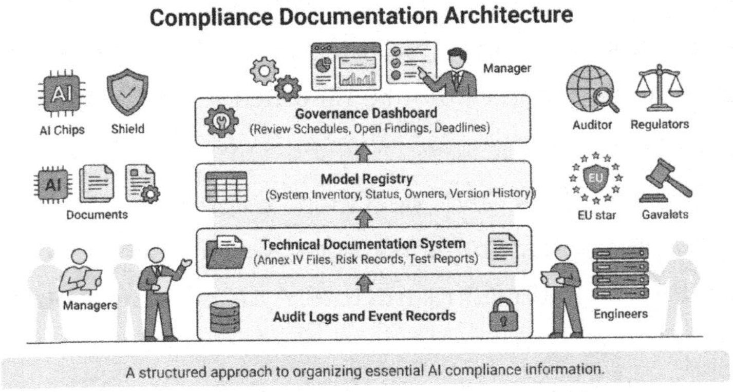

Automated Monitoring Tools for Post-Deployment Performance and Drift

Post-deployment monitoring is both a quality management practice and a regulatory obligation. For high-risk AI systems, the EU AI Act requires providers to establish a post-market monitoring plan that systematically collects and reviews performance data from deployed systems. The monitoring plan must be proportionate to the nature of the system and its risks, and must enable the provider to detect and respond to systematic deficiencies in system performance.

Automated monitoring tools for AI systems typically address two categories of performance issues. The first is technical performance degradation: the model's accuracy, precision, recall, or other performance metrics deteriorate over time due to a distributional shift between training data and current input data. Technical performance monitoring can be implemented using standard MLOps tooling that tracks model output distributions and statistical metrics over time. The second category is compliance-relevant performance issues: the model's behavior in specific demographic segments or operational contexts diverges from its tested behavior in ways that create regulatory risk. Compliance-relevant monitoring requires more intentional configuration because the relevant metrics are defined by the specific system's compliance requirements, not by generic model quality measures.

Drift detection is a critical component of post-deployment monitoring for high-risk AI systems. Input data drift, where the distribution of inputs shifts away from the training distribution, and concept drift, where the relationship between inputs and the correct output changes over time, are both potential triggers for reassessing the system or retraining. Monitoring systems should be configured with drift thresholds that, when exceeded, trigger a review of the system's conformity status. These thresholds should be documented in the post-market monitoring plan as part of the technical file.

Runbooks for monitoring response are as important as the monitoring infrastructure itself. A runbook specifies the steps to take when a monitoring alert fires: how to investigate the alert, how to determine whether it represents a genuine performance issue or a monitoring artifact, what the remediation options are, how to document the investigation and resolution, and under what circumstances the alert should trigger the incident escalation process described earlier in this chapter. Without runbooks, monitoring alerts produce organizational noise rather than assurance of compliance.

Integrating Compliance Tooling into Existing Enterprise Technology Stacks

One of the most practically important decisions in AI compliance technology is how to integrate compliance tooling with existing enterprise systems. Most organizations already operate model development platforms, data management infrastructure, CI/CD pipelines, IT service management systems, and enterprise risk management platforms. The question is whether AI compliance tooling should be a separate system that must be manually synchronized with these existing systems, or whether compliance requirements can be embedded in the existing systems by extending them with compliance-specific functionality.

The first approach of integration is generally preferable for organizations with mature technology infrastructure. Embedding compliance gates in existing CI/CD pipelines ensures that compliance checks are executed every time a model is deployed or updated, without requiring developers to remember to run a separate compliance check. Extending the existing IT service management system to include AI-specific fields for risk classification and documentation status ensures that AI systems are tracked consistently with other IT assets. Connecting the model registry to the data catalog ensures that dataset provenance information is available when preparing technical documentation.

Purpose-built AI governance platforms offer a more integrated set of compliance tools but require their own integration with existing systems. They typically offer model inventory management, documentation templates aligned to regulatory requirements, audit workflow support, and monitoring dashboards. For organizations building AI compliance capability from scratch and lacking mature existing systems to extend, a purpose-built platform may be the faster path to compliance infrastructure. For organizations with mature existing systems, extending those systems is usually more sustainable because it avoids creating a parallel governance system that may fall out of sync with operational reality.

The pilot-to-production pathway is the critical test of whether compliance tooling is genuinely integrated. A compliance infrastructure that works well for a single team or a single system during a pilot phase, but cannot scale to dozens of systems across multiple business units, has not solved the problem. Organizations should design and evaluate their compliance tooling with the full production deployment scale in mind, not with a pilot-scale. This includes assessing whether the tooling can handle the volume of systems, the variety of system types, the geographic distribution of teams, and the audit trail requirements of a mature production AI operation.

The scaling dimension of compliance governance warrants particular attention because the EU AI Act's requirements do not diminish as an organization's AI footprint grows. Every new high-risk system added to the portfolio requires its own technical documentation, risk management records, and monitoring configuration. Organizations that design their compliance infrastructure to scale linearly with the number of systems will find that compliance becomes operationally unmanageable as the portfolio grows. Sustainable design is one in which much of the compliance work is automated or templated, shared infrastructure reduces per-system overhead, and the governance team's attention is focused on genuine judgment calls rather than routine documentation production.

Diagram 4.10 - Compliance Tooling Integration Architecture

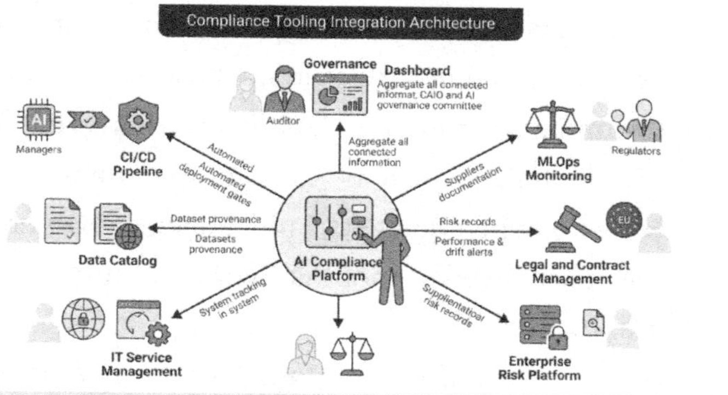

Streamlining AI Governance Through Integrated Enterprise Systems.

6.6 Manager's Checklist

Use this checklist to assess the maturity of your organization's AI compliance discipline. Each item corresponds to a concrete artifact or operational behavior, not a policy statement. If you cannot point to a specific artifact or demonstrate a specific behavior, the item is not yet complete.

- Each AI system in the organization's portfolio has a completed risk classification record that documents the classification determination and its reasoning, not merely the conclusion.
- The AI governance committee has formal terms of reference, a defined membership with specific roles, a meeting cadence, and documented records of decisions made (approval records, deviation records, and escalation dispositions).
- A model registry exists that tracks every AI system across its full lifecycle, including risk classification status, conformity assessment status, owner, version history, and post-market monitoring configuration.
- Pre-deployment review checklists are role-specific, concrete enough to be applied consistently by different

reviewers, and have been used for every high-risk system placed in service in the past twelve months.

- Compliance training has been delivered in role-differentiated formats to data scientists, engineers, procurement managers, and business owners within the past twelve months, with completion records maintained.
- Incident detection monitoring is configured for every deployed high-risk AI system with compliance-relevant thresholds documented in the post-market monitoring plan, and runbooks exist for investigating and responding to alerts.
- AI supplier contracts for procured AI components contain specific provisions covering documentation rights, modification notification, regulatory cooperation, and documentation retention.
- An escalation protocol is documented and has been tested in a tabletop exercise within the past twelve months, with findings addressed in a written remediation plan.
- Technical documentation for all in-scope AI systems is stored in a controlled system with access controls, version history, and a documented review schedule for keeping documentation current.
- The compliance technology stack is integrated with existing CI/CD, data management, and IT service management systems so that compliance checks occur as part of normal operational workflows rather than as separate manual processes.

6.7 Takeaway

Compliance as an organizational discipline means that every aspect of how the organization builds, procures, and operates AI systems reflects the requirements of the regulatory environment. It means that compliance evidence is generated naturally by the people doing the work, that accountability is explicit and tested rather than assumed, that governance processes are documented and consistently followed, and that technology infrastructure supports rather than burdens the people responsible for compliance. This is not a description of where most organizations are today; it is a description of where the EU AI Act requires them to be.

The path from policy-heavy compliance to genuine operational discipline is iterative. Organizations do not achieve it in a single program. They achieve it by identifying the highest-value gaps between their current state and the required state, addressing those gaps systematically, measuring the results, and repeating. Managers who treat the three pillars as a diagnostic framework, not a one-time project, and who build a culture where compliance is understood as mission-aligned operational quality rather than a regulatory burden, will build the kind of organization that regulators recognize and competitors cannot easily replicate.

7 The Conformity Assessment Imperative

A regulatory affairs director at a healthcare technology company described the first time her team tried to prepare a conformity assessment package for their AI-assisted diagnostic imaging tool as 'reverse-engineering what we should have been doing all along.' The development team had built a technically sophisticated system. It had been trained on a carefully curated dataset, had achieved strong performance benchmarks, and had been reviewed internally by the clinical team before deployment. But when the regulatory affairs team began assembling the technical documentation required under Annex IV of the EU AI Act, they discovered that the records they needed had never been created. The data governance documentation did not exist in the form required. The risk management activities were conducted informally and were not recorded. The human oversight mechanism was described in a product roadmap slide rather than in a tested operational procedure. The company had built a good AI system by its own engineering standards; it had not built a compliant one by the Act's standards.

That experience is not unusual. Most organizations that are encountering the EU AI Act's conformity assessment requirements for the first time discover that the gap between their current documentation practices and the Act's requirements is wider than they expected, not because they have been negligent, but because the Act's requirements reflect a systematically different approach to evidence of trustworthiness than the software industry has historically used. Conformity assessment asks not whether the system works well in general, but whether every specific requirement under the Act for which providers are accountable has been met, documented, and verified. Preparing for that assessment is a discipline that must be built deliberately.

7.1 Why It Matters

The conformity assessment process is the Act's primary mechanism for ensuring that high-risk AI systems meet the Act's substantive requirements before reaching the market or being put into service. It converts regulatory requirements into a concrete, verifiable demonstration that a specific system, in its current configuration, satisfies the applicable standards. For managers, this means that conformity assessment is not a post-development formality; it is a design constraint that must inform how AI systems are built, how data is documented, how testing is conducted, and how oversight mechanisms are configured from the earliest stages of development.

The assessment also has direct market consequences. High-risk AI systems subject to the Act cannot lawfully be placed on the EU market or put into service in the EU without a completed conformity assessment and, where applicable, CE marking. For organizations that sell AI-enabled products or services in the EU, conformity assessment is a market access requirement with the same commercial significance as any other regulatory certification in a regulated industry. Organizations with experience in medical device CE marking, aviation certification, or financial services regulatory approval will recognize the pattern; for those that do not, building the capability requires dedicated attention and lead time.

Beyond the legal requirement, conformity assessment capability creates operational value. Organizations that build systematic assessment processes accumulate structured knowledge about their AI systems that improves their ability to manage risk, identify vulnerabilities, and respond to incidents. The technical documentation required by the Act, when genuinely maintained, is a comprehensive reference for every person who works on or with the system, including operations teams, support teams, auditors, and successor development teams. Organizations that treat conformity assessment as a burden to minimize are forgoing a capability that pays dividends well beyond regulatory compliance.

7.2 The Architecture of Conformity Assessment

The EU AI Act establishes a structured framework for demonstrating that high-risk AI systems comply with its requirements. Understanding that framework at a structural level is a prerequisite to implementing it effectively. The framework's key design choices, including who performs the assessment, what standards apply, and what the CE mark signifies, each have practical implications for how organizations plan and execute their compliance programs.

Self-Assessment vs. Third-Party Assessment: When Each Pathway Applies

The EU AI Act provides two conformity assessment pathways for high-risk AI systems. The first and more widely applicable pathway is self-assessment, under which the provider conducts the conformity assessment using its own internal processes and personnel. The self-assessment pathway is available for most high-risk AI systems listed in Annex III of the Act. It does not mean that the assessment is less rigorous; it means that the provider bears full responsibility for conducting and documenting the assessment correctly, without a third party validating the work.

The second pathway involves a third-party assessment by a notified body, an independent conformity assessment organization accredited by a national authority to assess specific types of AI systems. Third-party assessment is required in specific circumstances: for AI systems intended for use as safety components in products covered by existing product safety directives where those directives require third-party conformity assessment; for biometric identification systems where the Act specifically requires notified body involvement; and in other cases where the Act or delegated acts specifically mandate external assessment. The provider cannot choose to use self-assessment for these systems.

For most enterprise AI deployments, the practical question is whether the specific system falls into a category that requires the involvement of a notified body. This determination requires careful reading of the Act's provisions and the product safety directives that may apply to the AI system or the product in which it is embedded. For AI systems that are standalone software products not embedded in physical products covered by existing safety directives, self-assessment is typically the applicable pathway. For AI systems embedded in medical devices, machinery, or other products subject to existing EU product safety regulations, the applicable pathway depends on the requirements of the relevant product directive.

Both pathways require the same underlying technical documentation. The difference between self-assessment and third-party assessment is not in the content of the technical file, but in who reviews it, how rigorously the review is documented, and who signs the declaration of conformity. Under self-assessment, the provider's authorized representative signs the declaration. Under a third-party assessment, the notified body issues a certificate that the provider includes in the declaration. In both cases, the provider retains legal responsibility for the accuracy of the conformity claim.

Organizations planning their conformity assessment strategy should make the self-assessment versus third-party determination early, because it has significant resource implications. Third-party assessment by a notified body requires identifying an accredited body with capacity and relevant scope, scheduling assessment activities that can have long lead times, preparing documentation to the specific standards the assessor applies, and budgeting for assessment fees that can be substantial for complex systems. These requirements need to be factored into product development timelines, not discovered when the system is ready for market.

Diagram 5.1 - Conformity Assessment Pathway Decision

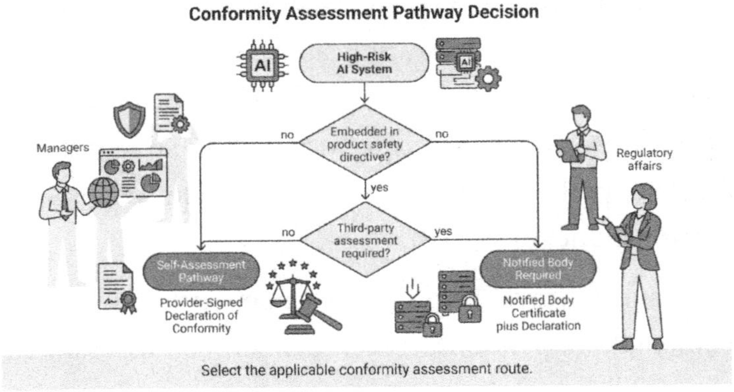

The Role of Harmonized Standards in Defining Assessment Criteria

One of the most practically significant tools in the conformity assessment process is the harmonized standard. A harmonized standard is a technical standard, typically developed by a European standards organization such as CEN or CENELEC, that the European Commission has formally recognized as meeting specific provisions of a regulation. When an AI provider demonstrates that its system complies with the relevant harmonized standards, it benefits from a presumption of conformity: it is presumed to comply with the regulatory requirements covered by those standards unless evidence to the contrary is presented.

The presumption of conformity is valuable for two practical reasons. First, it provides a defined, externally validated set of criteria that the provider can design and test against, rather than having to interpret the Act's more general requirements independently. Second, it creates a clearer audit trail because the provider can document compliance with specific standard clauses rather than making more general assertions about satisfying regulatory requirements. Auditors and notified bodies understand compliance with harmonized standards better than they understand ad hoc compliance narratives.

For AI systems specifically, the harmonized standards landscape is still developing. The EU AI Act was adopted in 2024, and the development of harmonized AI-specific standards is an ongoing process by the relevant standards bodies. Organizations working toward conformity assessment today must therefore work with a combination of existing harmonized standards from adjacent domains, such as software quality standards, functional safety standards, and information security standards, and emerging AI-specific standards that may not yet have formal harmonized status. The CEN-CENELEC Joint Technical Committee 21 has been tasked with developing AI-specific standards for use under the Act, and tracking the progress of that work is part of a well-designed regulatory monitoring program.

When harmonized standards are not available or do not cover all relevant aspects of a system, providers must demonstrate conformity through alternative means, such as their own technical specifications, industry best practices, or detailed documentation of how they interpret and apply the Act's requirements. This is a more demanding position because it requires the provider to make and defend their own interpretive judgments about what the Act requires, rather than relying on the presumption of conformity provided by a standard. Organizations should monitor the harmonized standards pipeline closely and update their assessment approaches as relevant standards achieve harmonized status.

Diagram 5.2 - Harmonized Standards and Conformity Presumption

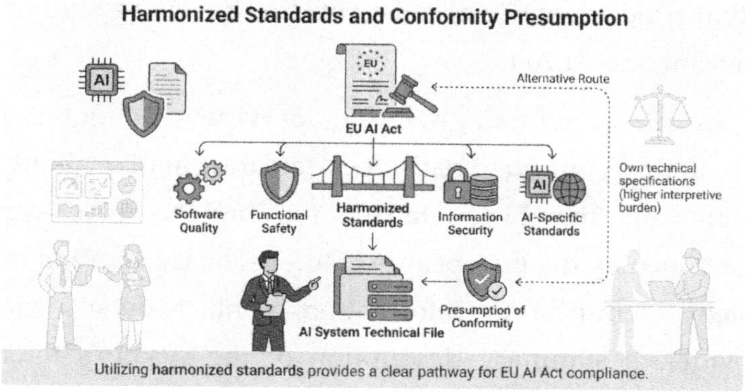

CE Marking and the EU Database for High-Risk AI Systems

The CE marking is the visible output of a completed conformity assessment. It is a declaration that the product meets the essential requirements of all applicable EU legislation, as required by the manufacturer or, for products from outside the EU, by the authorized representative. For high-risk AI systems, affixing the CE mark is not a formality; it is a legal statement that the provider has completed the conformity assessment process, has compiled the technical documentation, and has signed the EU declaration of conformity.

The mechanics of CE marking for standalone AI software products differ from those for AI systems embedded in physical products. For standalone AI software, the CE mark is applied to the software itself, typically in accompanying documentation, the user interface, or the distribution packaging, because software does not have a physical surface. For AI systems embedded in physical products, the CE mark is affixed to the product. It covers all requirements applicable to that product under all relevant directives and regulations, including the AI Act.

In addition to CE marking, providers of high-risk AI systems are required to register certain information about their systems in the EU database for high-risk AI systems maintained by the European AI Office. The registration covers system identification information, contact details for the provider, a summary description of the system's intended purpose, and the conformity assessment status. The database is publicly accessible for most system categories, creating a layer of transparency that allows regulators, deployers, and the public to identify registered high-risk AI systems. Providers should build database registration into their conformity assessment timeline as a required step, occurring before or at the time of placing the system on the market.

Market surveillance authorities use the database as one of their monitoring tools, cross-referencing registered systems with complaints, incident reports, and their own proactive surveillance activities. Organizations should therefore treat the database registration as a compliance-sensitive artifact whose accuracy must be maintained over time: updates to system capabilities, substantial modifications, and changes in conformity assessment status must be reflected in the registration promptly.

Diagram 5.3 - CE Marking and EU Database Registration Flow

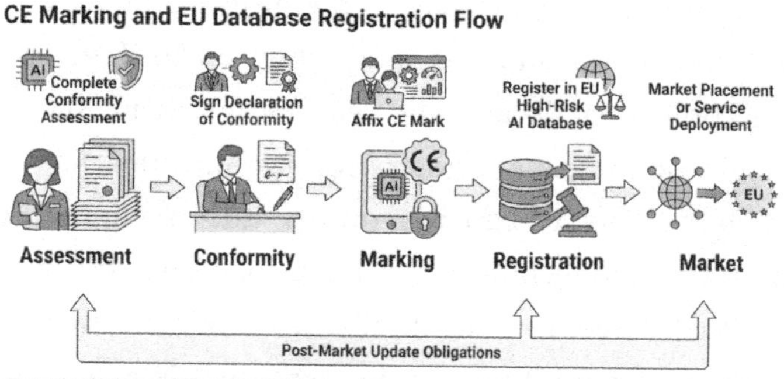

Demonstrating compliance for high-risk AI systems entering the EU market.

7.3 Technical Documentation Requirements

The technical documentation required under the EU AI Act is set out in Annex IV to the Act. It is comprehensive by design: the EU legislator made a deliberate choice to require providers to compile structured, verifiable evidence of how their AI systems were designed, trained, tested, and validated, rather than relying on assertions of quality. For managers, this means that producing the technical file is not a documentation exercise that happens at the end of development; it is a discipline that must be embedded throughout development.

What the Act Requires in the Technical File

Annex IV of the EU AI Act specifies nine categories of information that the technical documentation must contain. The first is a general description of the AI system, including its intended purpose, the version information, and the hardware and software environment in which it is designed to operate. This description must be detailed enough for a reviewer unfamiliar with the system to understand what it does and what it is designed to do.

The second category covers the design and development of the system in detail: the methods used to develop it, the design choices made and the reasons for them, the general logic and algorithms underlying it, and the key design parameters. This category requires documentation of the reasoning behind technical decisions, not merely the outcomes. Teams that have not habitually documented their design rationale will find this requirement requires a change in practice, not just an accumulation of existing documents.

The third category addresses training, testing, and validation. It requires documentation of the training data used, including its provenance, selection criteria, data collection methods, and preprocessing steps. It requires documentation of the testing procedures used and the results obtained, including performance metrics across relevant operational conditions and subpopulations. It requires documentation of any known system limitations and the circumstances under which performance may fall below acceptable levels.

The fourth through ninth categories cover risk management documentation (per Article 9 requirements), human oversight measures, accuracy, robustness, and cybersecurity specifications; quality management documentation; the declaration of conformity; and the instructions for use. The instructions for use are a particularly operationally important element because they govern how deployers are expected to configure and operate the system, and they define the boundaries of the provider's conformity claim. A deployer who operates a system outside the boundaries specified in the instructions for use assumes liability for any resulting non-conformity.

The complete technical file for a complex high-risk AI system can be a substantial body of documentation, potentially spanning hundreds of pages across multiple documents. Organizations should develop a technical file template and a documentation management process that systematically organize the required content, maintain version control, and enable updating individual sections without disrupting the file's overall coherence. The template should map each documentation requirement to the Annex IV item it satisfies, enabling efficient assessment of completeness.

Data governance documentation in the technical file deserves particular emphasis because it is one of the most demanding and most frequently incomplete elements of the conformity assessment preparation process. The Act requires documentation of the characteristics and appropriateness of the datasets used for training, validation, and testing, including statistical properties, potential biases, and steps taken to address those biases. For most organizations, this requires close collaboration between data engineering teams, who have the technical information about the datasets, and compliance teams, who must translate that information into documentation that satisfies the Act's requirements.

Diagram 5.4 - Annex IV Technical File Structure

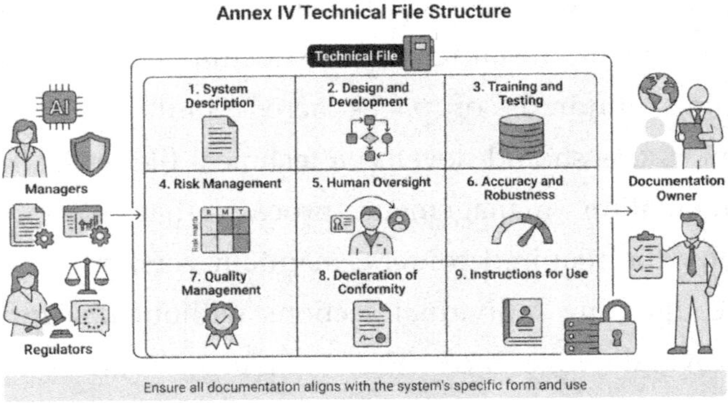

Logging and Record-Keeping Obligations During and After Deployment

The EU AI Act imposes logging requirements on high-risk AI systems that go beyond what most software organizations have historically maintained. High-risk AI systems must be designed to automatically generate logs that record the system's operation throughout its lifecycle, and, to the extent technically feasible, to record events that could affect system performance and serious incidents. The deployer must retain these logs for at least six months, and longer where sector-specific regulations require it.

The logging requirement has two distinct purposes. The first is prospective: logs enable the post-market monitoring required by the Act, allowing providers and deployers to identify patterns in system behavior over time that might indicate performance degradation, systematic errors, or emerging risks. The second is retrospective: in the event of an incident or a supervisory authority investigation, logs are the primary evidence for reconstructing what the system did, when, and in response to what inputs. The investigative value of logs is directly proportional to the completeness and integrity of the logging design.

Designing logging capability to meet Act requirements requires decisions at the system architecture level. What events will be logged? At what granularity will inputs and outputs be recorded? How will logs be stored securely to prevent modification while remaining retrievable for the required retention period? How will logs be structured to enable efficient search and analysis by compliance and incident investigation teams? These decisions must be made before system deployment and documented in the technical file.

Privacy considerations add complexity to AI logging design, particularly for systems that process personal data. The inputs to many high-risk AI systems include personal data about the individuals whose cases the systems process. Logging at the input level, therefore, involves recording personal data. Organizations must ensure that their logging design complies with the GDPR's data minimization and purpose limitation principles while also satisfying the Act's logging requirements. Where these requirements create tension, the organization should document its resolution approach in the technical file and seek legal advice on the appropriate balance.

Record-keeping obligations extend beyond operational logs to the full set of compliance documentation. The Act requires providers to keep the technical documentation for ten years after the last AI system of a given model was placed on the market or put into service. For products with long market lives, this can mean maintaining documentation for systems that are no longer actively developed or supported. Organizations should build documentation, archiving, and retention management into their compliance infrastructure from the beginning, rather than treating them as afterthoughts when systems reach the end of life.

Diagram 5.5 - AI System Logging and Retention Requirements

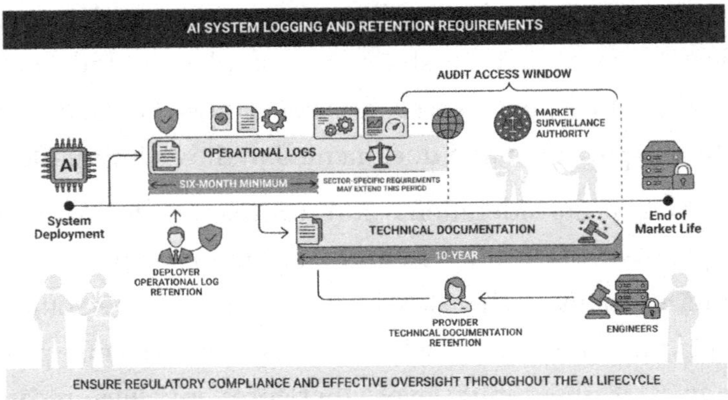

Documentation as Living Evidence: Update Obligations and Version Control

One of the most important conceptual shifts that AI Act compliance requires from many organizations is treating technical documentation as a living system rather than a project deliverable. In typical software development practice, documentation is produced at the end of a development cycle and then becomes stale as the system evolves. Under the EU AI Act, technical documentation must remain current and accurate throughout the system's market life. If the system is modified in a way that affects its conformity, the documentation must be updated and the conformity assessment reviewed.

The substantial modification rule is the key trigger for documentation updates and conformity re-assessment. The Act defines a substantial modification as a change that affects the AI system's compliance with the Act or results in a change to the intended purpose for which the system has been assessed. Determining whether a specific change constitutes a substantial modification requires judgment: a change to the user interface that does not affect the system's decision logic may not be substantial, while a change to the training data selection criteria or the model architecture may well be. Organizations should document their substantial modification determination process in their quality management system and apply it consistently to every change that reaches the decision gate.

Version control for technical documentation should mirror the version control practices used for the software system itself. Each version of the system should have a corresponding technical documentation version that accurately describes it. The documentation version history should be maintained and accessible, so that at any point in time, it is possible to retrieve the documentation describing the system in a particular version as of a particular date. This capability is essential for investigating incidents that may have occurred during the operation of an earlier version.

Documentation review cycles should be built into the system's operational governance calendar. Even in the absence of modifications, technical documentation should be reviewed periodically, at least annually, to verify that it remains accurate given any changes in the operational environment, changes in the understanding of the system's performance characteristics, or changes in the regulatory or standards landscape that affect its conformity assessment. The review should be conducted by the technical documentation owner and produce a documented conclusion: either the documentation is current and accurate, or specific updates are required, with a timeline for their implementation.

7.4 Third-Party Auditing and Notified Bodies

For organizations whose AI systems require third-party conformity assessment, engagement with a notified body is a significant operational undertaking. Notified bodies are independent conformity assessment organizations that have been formally accredited and authorized by national authorities within EU member states to conduct assessments for specific regulatory frameworks. Understanding how the notified body system works and how to prepare for and manage a third-party assessment is essential for organizations in affected categories.

The Notified Body System: Accreditation, Scope, and Market Capacity

Notified bodies derive their authority from a dual accreditation process: they must first be accredited by a national accreditation body to demonstrate that they meet the competence and impartiality requirements for conformity assessment, and they must then be formally notified by their national authority to the European Commission, which records them in the NANDO database and specifies the scope of regulations and system categories they are authorized to assess. Not all notified bodies are authorized to assess AI systems, and among those that are, their scope may be limited to specific AI system categories.

The market capacity question is significant for organizations planning their conformity assessment timelines. The notified body system for AI is newer than for established regulated industries such as medical devices or pressure equipment, and the number of accredited notified bodies with AI-specific scope is limited relative to the anticipated volume of systems that will require third-party assessment as the Act's requirements phase in. Organizations that require notified body assessment should begin engaging with potential notified bodies early, assess their capacity and lead times, and factor assessment scheduling into their product launch timelines.

Selecting a notified body requires consideration of several factors beyond accreditation status. The body's technical expertise in the relevant AI application domain is significant to the quality and efficiency of the assessment. A notified body with deep expertise in healthcare AI will conduct a more technically informed review of a medical imaging AI system than a body whose expertise is primarily in manufacturing. Assessment fees, turnaround times, and the body's experience with the specific standards applicable to the system type are also relevant selection criteria. Organizations should conduct a structured selection process and document the rationale for their selection.

The relationship with a notified body is not transactional but ongoing. Once a notified body has assessed a system and issued a certificate, it must be notified of substantial modifications to the system that may affect the certificate's validity. The body may conduct surveillance assessments during the certificate period to verify that the provider's quality management system and technical documentation continue to meet the required standards. Organizations should build the cost and operational requirements of this ongoing relationship into their product operational budgets from the outset.

Diagram 5.6 - Notified Body Accreditation and Selection

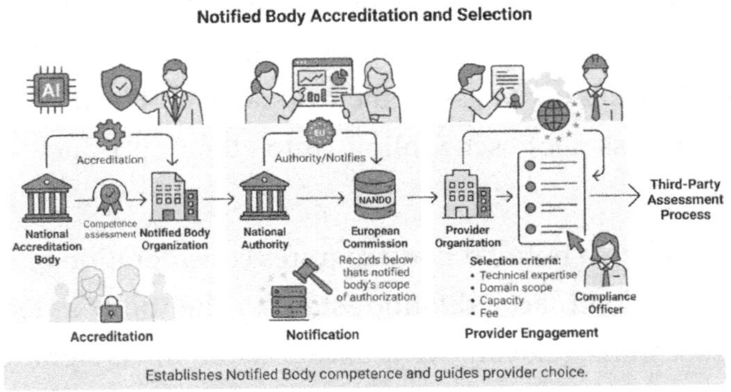

Preparing for a Third-Party Audit: What Assessors Look For

A third-party conformity assessment is not a surprise inspection. The provider knows in advance what the assessment will cover because of the Act's requirements, and the applicable standards specify the assessment criteria. The assessment will evaluate whether the technical documentation is complete, accurate, and consistent with the actual system; whether the provider's quality management system meets the Act's requirements; whether the risk management activities have been properly conducted and documented; and whether the system has been tested and validated according to the applicable requirements.

The most common assessment findings that delay or prevent certification fall into predictable categories. Documentation incompleteness is the most frequent issue: the technical file exists, but specific Annex IV elements are missing or insufficiently detailed. Inconsistency between documentation and the actual system is the second most common finding: the documentation describes design choices or performance characteristics that do not reflect the system as actually built and operated. Process gaps are the third category: the provider cannot demonstrate that their quality management system was actually applied to the development of this specific system, rather than existing only on paper.

Preparation for a third-party assessment should begin with a structured internal pre-assessment. The internal pre-assessment should systematically evaluate each element of the technical documentation against the Act's requirements and applicable standards, identify gaps, assign remediation owners, and track completion before scheduling the external assessment. Many organizations engage external legal or compliance advisors with conformity-assessment experience to support the internal pre-assessment, because in-house teams may have blind spots to gaps that are obvious to experienced external reviewers.

The practical logistics of a third-party assessment require planning. The provider must provide the notified body with access to the technical documentation, typically in digital form, in advance of the assessment. The assessment itself typically involves document review, technical discussions with the development team and quality management representatives, and potentially practical testing of the system. Assessment timelines can range from weeks to months, depending on the system's complexity and the notified body's workload. Organizations should not compress the preparation timeline and expect the assessment to be completed quickly; the quality of preparation directly determines the assessment's efficiency.

Diagram 5.7 - Third-Party Assessment Preparation and Execution

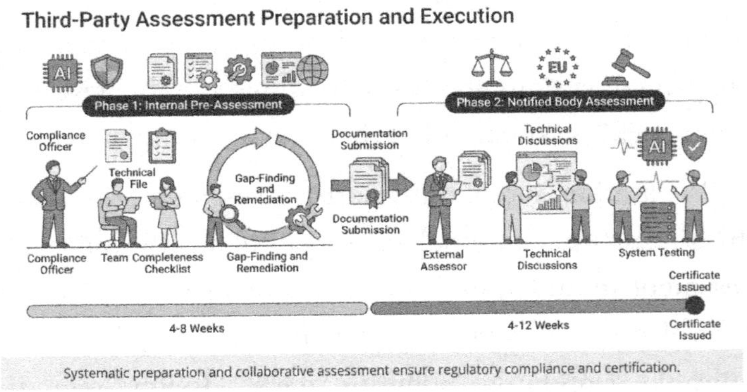

Managing Ongoing Relationships with Auditors and Assessment Timelines

Conformity assessment is not a one-time event, but a relationship managed over the system's lifetime. Once a certificate is issued, the provider enters an ongoing relationship with the notified body that includes obligations to notify the body of substantial modifications, to cooperate with surveillance assessments during the certificate period, and to apply for reassessment if the system undergoes changes that affect the certificate's scope.

Managing this relationship effectively requires assigning clear internal ownership of the notified body relationship, maintaining a communication log of all interactions with the notified body, and tracking the certificate period timeline and upcoming reassessment milestones. The technical documentation owner is typically the most appropriate owner of the notified body relationship because they have the deepest knowledge of the system's documentation and are best positioned to respond to the body's technical questions.

Organizations with multiple AI systems requiring third-party assessment must also manage the aggregate timeline across their portfolio. Different systems may have certificates with different issuance dates and different validity periods, creating a complex mosaic of renewal and surveillance milestones that must be tracked proactively. A portfolio-level assessment calendar, maintained as part of the model registry or the compliance management system, is an essential tool for preventing certification gaps that would require systems to be withdrawn from the market pending reassessment.

7.5 Building the Internal Conformity Assessment Capability

Organizations with a single high-risk AI system can approach conformity assessment as a one-time project. Organizations with multiple high-risk AI systems, or with active AI development pipelines that will regularly produce new high-risk systems, must build conformity assessment as an internal capability: a repeatable process that can be applied systematically to each new system as it approaches market readiness, staffed by people who understand the requirements, and supported by tools and templates that ensure consistency.

Establishing a Repeatable Assessment Process for the AI Portfolio

A repeatable conformity assessment process begins with standardization: a defined set of steps that every new AI system in scope must progress through from initial development through market placement, with defined inputs and outputs at each step, defined roles responsible for each step, and a defined quality check that each step's output is complete before the next step begins.

The process design should begin with a gap analysis between the organization's current development and documentation practices and the Act's requirements. The gap analysis identifies which required activities are already part of the development process (even if not currently documented in the format required), which activities need to be added, and which documentation outputs need to be created or restructured. The process design then embeds the new activities and documentation requirements into the existing development lifecycle at the appropriate points.

Templating is one of the highest-leverage tools for building assessment efficiency. A comprehensive set of templates for each section of the technical file, pre-populated with the Act's structural requirements and guidance notes explaining what content each field requires, dramatically reduces the effort required to produce compliant documentation for a new system. Templates also improve consistency across systems and among team members with varying levels of familiarity with the Act's specific requirements. The development of good templates requires significant upfront investment but pays dividends on every subsequent system.

The assessment process should include a staged review protocol: an internal completeness review after the initial draft of each technical file section is produced, a cross-functional technical review involving data science, engineering, legal, and risk stakeholders before the complete file is finalized, and a final sign-off review by the technical documentation owner and the CAIO before the declaration of conformity is signed. Each review stage should have a defined scope, a defined set of reviewers, and a defined output (a review record with findings and their dispositions).

Process metrics are essential for managing assessment capability as the portfolio grows. Key metrics include the average time from assessment initiation to completion of the technical file, the number and type of findings identified at each review stage, the rate of findings that are new versus recurrences of previously identified gaps, and the time from identification of substantial modifications to documentation update. These metrics identify where the process is working well and where additional training, tooling, or process redesign is needed.

Diagram 5.8 - Repeatable Conformity Assessment Process

Repeatable Conformity Assessment Process

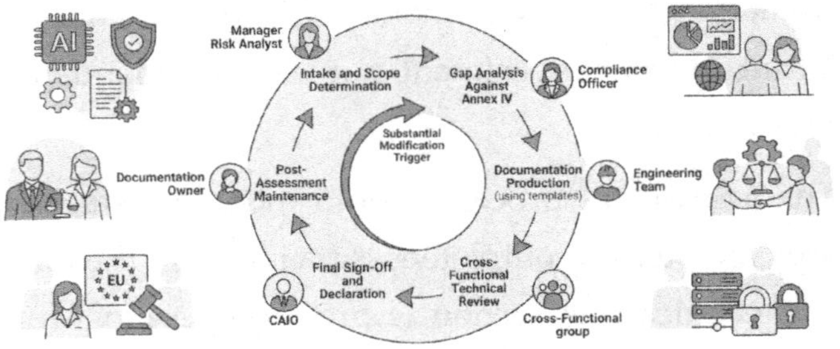

A structured approach for continuous compliance across the AI lifecycle.

The Intersection of Conformity Assessment and Product Development Cycles

The most persistent tension in conformity assessment program design is the conflict between the documentation-intensive, carefully sequenced nature of conformity assessment and the iterative, fast-moving nature of AI development. Modern AI development commonly uses agile or continuous delivery methodologies designed for rapid iteration, whereas conformity assessment was designed for a more sequential development model. Reconciling these approaches requires deliberate process design.

The key design principle is that conformity assessment activities must be integrated into the development cycle at the points where the underlying development activities occur, not deferred to a separate compliance phase after development is complete. Data governance documentation should be produced during the data preparation phase of the development process. Risk management activities should be conducted in parallel with design and architecture decisions, not after them. Testing documentation should be compiled as testing is executed. Human oversight mechanisms should be designed during system architecture and tested during system validation, not specified in a document after the system is already built.

Sprint-level compliance integration means identifying which conformity assessment requirements can be addressed within the normal development sprint cycle and assigning them to the sprint, along with the corresponding development work. Not all requirements can be addressed at the sprint level: the final compilation and review of the complete technical file, the declaration of conformity, and the CE marking process require a distinct project phase after the iterative development is complete. But the underlying content that populates the technical file can and should be generated during development, which dramatically reduces the time and effort required for the final compliance phase.

Release cadences for AI systems in active development must account for the documentation update and re-assessment obligations that each release may trigger. Organizations that release AI model updates on a frequent cadence need a defined process for evaluating each release against the substantial modification criteria and updating documentation and conformity status accordingly. Releasing updates without this assessment creates conformity risk that accumulates silently and is discovered only during an audit or an incident investigation.

Diagram 5.9 - Conformity Assessment in Agile Development Cycles

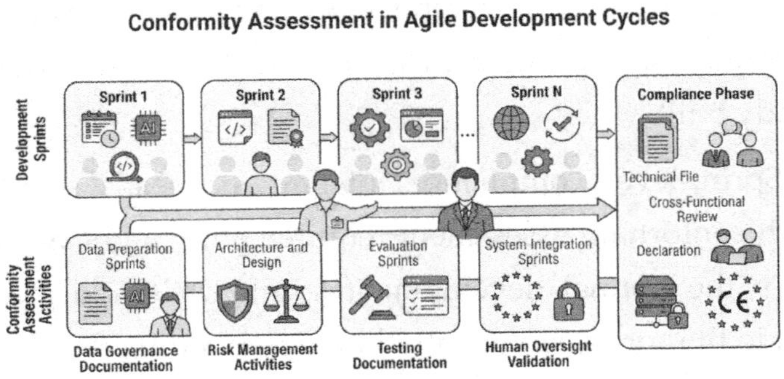

Integrating continuous compliance and iterative development for efficient regulatory alignment.

Lessons from Medical Devices and Aviation on Systematic Compliance Assurance

Two regulated industries have operated systematic conformity assessment programs under EU regulation for decades and offer directly applicable lessons for AI compliance program design. The medical device industry has operated under the Medical Devices Regulation (MDR) and the In Vitro Diagnostic Regulation (IVDR), both of which impose rigorous technical documentation, quality management, and post-market surveillance requirements, with a structure similar to the EU AI Act. The aviation industry operates under EASA certification frameworks that require extensive safety case documentation and systematic safety assessment for new aircraft systems and components.

The most important lesson from the medical device industry is the value of building quality management systems that generate compliance documentation as a natural byproduct of normal quality management activities, rather than treating compliance documentation as a separate workstream. Medical device manufacturers with mature quality management systems can produce conformity assessment packages efficiently because their design controls, risk management processes, and testing protocols are already structured to generate the required evidence. Organizations building AI compliance capability should design their quality management systems with the same philosophy: make compliant documentation the natural output of well-executed development and quality activities.

The aviation industry offers a different but equally important lesson: the value of explicit safety cases and design assurance evidence. Aviation certification requires developers to demonstrate, through structured argumentation supported by evidence, that their systems meet safety objectives. This approach, adapted to AI, produces more rigorous and more auditable conformity evidence than simple compliance checklists, because it requires the provider to explain not just what they did but why what they did is sufficient to meet the requirement. Organizations building conformity assessment capabilities for complex, novel AI systems should explore structured assurance case approaches as a complement to the Act's documentation requirements.

Both industries also share a lesson about the importance of regulatory relationship management: organizations that engage proactively with regulators, seek interpretive guidance early on novel questions, and demonstrate good faith in their compliance efforts receive more constructive treatment in audit and enforcement contexts than organizations that minimize their regulatory interactions and treat compliance as a purely legalistic exercise. The EU AI Act is new enough that many interpretive questions remain genuinely open, and organizations that engage constructively with the supervisory authority and the European AI Office on those questions will shape the emerging compliance expectations in ways that benefit the entire industry.

Diagram 5.10 - Cross-Industry Compliance Assurance Lessons

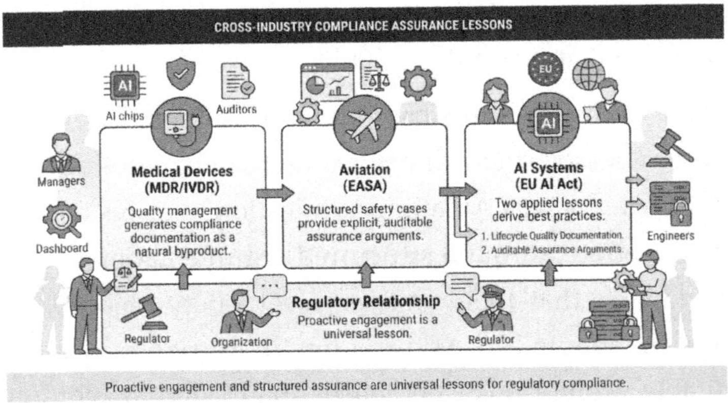

7.6 Manager's Checklist

Use this checklist to assess your organization's readiness for conformity assessment. Each item reflects a concrete capability or artifact rather than a policy statement. Assess each item against your highest-risk AI systems currently in development or deployed.

- Every high-risk AI system in development or deployment has a documented conformity assessment pathway determination, specifying whether self-assessment or notified body assessment applies, with the reasoning recorded.
- A complete technical file template aligned to Annex IV has been developed and is in use for all new high-risk AI system documentation efforts.

- Data governance documentation for each high-risk AI system covers training, validation, and testing datasets, including provenance, selection criteria, known biases, and mitigation steps.
- A substantial modification determination process exists, is documented in the quality management system, and has been applied to every system update in the past twelve months with a written determination record.
- Logging and monitoring are configured for every deployed high-risk AI system with retention periods meeting the six-month minimum, and log retrieval has been tested against an incident investigation scenario.
- For systems requiring notified body assessment, a qualified notified body has been identified and engaged, with an assessment timeline integrated into the product launch plan.
- A technical documentation review cycle is scheduled for all in-scope AI systems at least annually, with a documented review record confirming currency or specifying required updates.
- Internal pre-assessment procedures are documented and have been used for every high-risk system that has undergone conformity assessment, with findings and their dispositions recorded.
- The EU database registration for all in-scope high-risk AI systems is current, accurate, and assigned to a named owner responsible for keeping it updated.

7.7 Takeaway

Conformity assessment is the moment where all of the EU AI Act's substantive requirements converge into a single, verifiable demonstration that a specific AI system meets the Act's standards. Organizations that approach it as a documentation compilation exercise discover too late that the documentation cannot be compiled credibly without the underlying development and quality management practices that generate the evidence. The sequence is not: build the system, then document it, then assess it. The sequence is: design for conformity, document as you go, compile and assess when complete.

The organizations that will manage conformity assessment most efficiently are those that build systematic assessment capability rather than managing each system as a separate compliance project. Templates, repeatable processes, integrated development workflows, and trained people who understand what the Act requires and why are the ingredients of a scalable assessment program. Those ingredients take time and deliberate investment to develop. Still, they deliver compounding returns as the AI portfolio grows and the Act's requirements become the permanent backdrop of AI product development in the EU market.

8 High Stakes, High Standards: Managing High-Risk AI

When a federal procurement officer realized that the automated decision-support tool her agency had been using for five years to assist with benefits eligibility determinations would be classified as a high-risk AI system under the EU AI Act's framework, her first reaction was not alarm but recognition. The system made recommendations that affected whether applicants received housing support. It had been deployed with minimal documentation, no formal risk management records, and a human review process that was more nominal than substantive: case workers routinely accepted the system's recommendations without independent verification because they were under time pressure and had been told the system was accurate. Nothing about that situation was unusual for public sector AI deployments of that era. Everything about it was incompatible with the Act's requirements.

That recognition moment, where leaders see their existing AI systems through the Act's requirements and find the gap wider than expected, is the starting point for high-risk AI management. The Act's high-risk categories cover precisely the domains where AI has been deployed most ambitiously and where the consequences of AI error fall most directly on individuals with limited recourse: hiring decisions, credit assessments, educational placements, law enforcement tools, migration determinations, and judicial support systems. The obligations the Act imposes for these systems are not bureaucratic additions to responsible development; they are the minimum infrastructure required to deploy AI in high-consequence domains with accountability.

8.1 Why It Matters

High-risk AI classification is the Act's primary mechanism for directing regulatory attention to the systems that pose the greatest risk of harm. The eight Annex III domains, covering biometrics, critical infrastructure, education, employment, essential services, law enforcement, migration, and the administration of justice, were selected because they represent areas where AI decisions directly affect fundamental rights, where the affected individuals often have limited ability to contest those decisions, and where the systemic effects of AI error can be severe and wide-reaching. When a manager identifies that their organization operates an AI system in one of these domains, the appropriate response is not to look for reasons the system might not be high-risk. It is to initiate the compliance process that the Act requires.

The obligations for high-risk AI systems are substantial but structured. The Act specifies what a quality management system must contain, what a risk management system must do, what data governance documentation must cover, what human oversight must look like, what accuracy and robustness standards apply, and what post-market monitoring must include. For managers who have operated in adjacent regulated domains, such as financial services or healthcare, many of these requirements will be familiar in structure, even if new in application to AI. For managers coming from less regulated environments, they represent a significant operational uplift that must be planned for explicitly.

The provider-deployer distinction is particularly consequential for high-risk AI management because it determines which obligations fall on which organizational role. Providers who develop and place high-risk AI systems on the market bear the primary responsibility for documentation, assessment, and quality management. Deployers who put those systems into use bear obligations, including human oversight, implementation of human oversight, log maintenance, fundamental rights impact assessment in certain circumstances, and compliance with the provider's instructions for use. Many organizations occupy both roles simultaneously, developing some AI systems and deploying others procured from external providers, and must manage both sets of obligations in their governance structures.

8.2 Classification: Identifying Your High-Risk Systems

The EU AI Act's high-risk classification system is not self-applying. It requires active analysis of each AI system against the Act's criteria, documentation of the analysis and its conclusion, and ongoing monitoring for changes in system characteristics or use that might alter the classification. Organizations that rely on informal impressions about whether their systems are high-risk create classification risk: the risk that a system is treated as non-high-risk when it actually meets the classification criteria, exposing the organization to enforcement action and liability.

The Annex III Categories: Reading the Classification Lists Accurately

Annex III of the EU AI Act lists eight domains in which AI systems are presumptively classified as high-risk. The first domain covers biometric identification and categorization: AI systems used for remote biometric identification of individuals in publicly accessible spaces, AI systems used for biometric categorization based on sensitive attributes such as race or political opinion, and AI systems used for emotion recognition. The biometric provisions are among the most politically significant in the Act, and their scope has been clarified by Commission guidance on the boundary between general biometric functionality and high-risk biometric identification.

The second domain covers critical infrastructure: AI systems intended to be used as safety components in the management and operation of critical digital infrastructure, road traffic, and the supply of water, gas, heating, and electricity. The key qualifier is 'safety component': an AI system that optimizes energy consumption in a data center may not be a safety component, whereas an AI system that controls load balancing in an electrical grid in a way that affects system stability is likely to be.

The third domain covers education and vocational training: AI systems intended to determine access to or assignment to educational institutions and vocational training programs, and to evaluate learning outcomes and assess educational performance. This covers automated admissions systems, performance assessment tools, and systems that make recommendations about educational pathways that have material consequences for individual students.

The fourth domain covers employment, worker management, and access to self-employment: AI systems used for recruitment and selection (including CV sorting, interview assessment, and promotion decisions), for work allocation and monitoring, and for evaluating employees' performance and behavior. This is one of the most commercially significant high-risk domains because AI-assisted hiring tools are widely deployed across industries, and many existing deployments have not been designed to the standard the Act requires.

The fifth domain covers essential private and public services: AI systems used by or on behalf of public authorities to evaluate eligibility for essential public benefits and services, AI systems used to evaluate creditworthiness, and AI systems used in life and health insurance risk assessment. This domain encompasses a wide range of financial services and public-sector AI applications that directly affect individuals' access to fundamental economic resources.

The sixth domain covers law enforcement: AI systems used by law enforcement authorities for individual risk assessment, polygraph-like testing, evaluation of evidence reliability, profiling-based predictive policing, crime analytics related to individuals, and crime or criminal recidivism prediction. The seventh domain covers migration, asylum, and border control: AI systems used for risk assessment of individuals arriving at borders, verification of travel documents, and assessment of asylum applications. The eighth domain covers the administration of justice and democratic processes: AI systems that assist courts in researching and interpreting facts and applying the law, and AI systems that influence elections and voting behavior.

Accurate reading of these categories requires attention to the system's intended purpose, not merely its technical capabilities. The Act classifies systems based on their designed and intended use. A general-purpose machine learning model is not automatically high-risk; its classification depends on what it is designed and intended to do. An AI system that performs facial recognition as a general technical capability but is not intended for use in law enforcement or border control may not be high-risk under the biometric domain provisions. However, if the system is subsequently put to a high-risk use by a deployer, the deployer may bear obligations arising from the actual use, even if the provider did not intend it.

Diagram 6.1 - Annex III High-Risk AI Domains

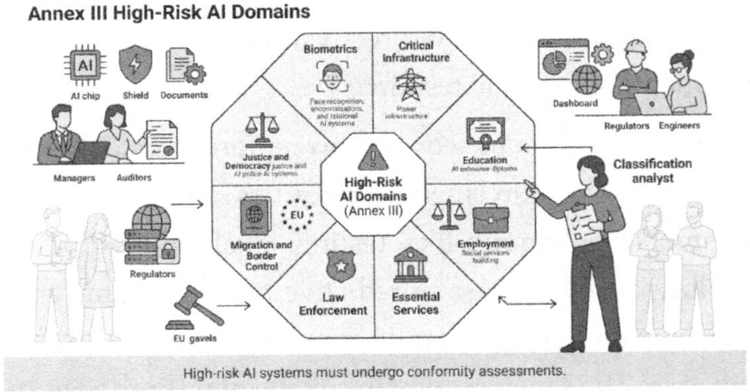

The Substantial Modification Rule and Reclassification Triggers

The EU AI Act's definition of substantial modification is the key boundary condition that determines when a high-risk

AI system must undergo a new conformity assessment. A substantial modification is a change to the AI system that affects the system's compliance with the Act or results in a change to the intended purpose for which the system has been assessed. This definition has two important implications. First, not all changes to an AI system are substantial modifications. Second, a change that affects a system's compliance is substantial, regardless of whether it was intended to do so.

In practice, the most common triggers for a substantial modification determination are changes to the system's intended purpose or scope of use, changes to the underlying model or algorithm that materially affect system performance, changes to the training data that affect the system's demographic behavior, changes to the human oversight mechanism, and changes to the system's inputs or output format that affect how decisions are made. Minor bug fixes, user interface improvements that do not affect system decision logic, and performance optimizations that do not affect the output distribution typically do not constitute substantial modifications. Still, each change should be assessed against the criteria rather than assumed to be non-substantial.

The practical governance challenge of the substantial modification rule is that AI systems, particularly those using machine learning, evolve continuously. Model retraining on new data, fine-tuning for new deployment contexts, and incremental feature additions can, individually, be non-substantial changes that, collectively, represent a substantially modified system. Organizations should establish a change accumulation tracking mechanism that monitors the cumulative effect of changes since the last conformity assessment and triggers a review when a defined threshold of accumulated change is reached, even if no individual change has cleared the substantial modification bar.

Reclassification can also be triggered by external changes rather than modifications to the system. A change in the regulatory definition of the intended purpose, a new Commission implementing act that extends the Annex III categories, or a deployer's decision to use the system in a new domain can all affect a system's classification status. Organizations should monitor the regulatory environment for classification-relevant developments and promptly apply them to their AI system inventory.

Diagram 6.2 - Substantial Modification Assessment Decision

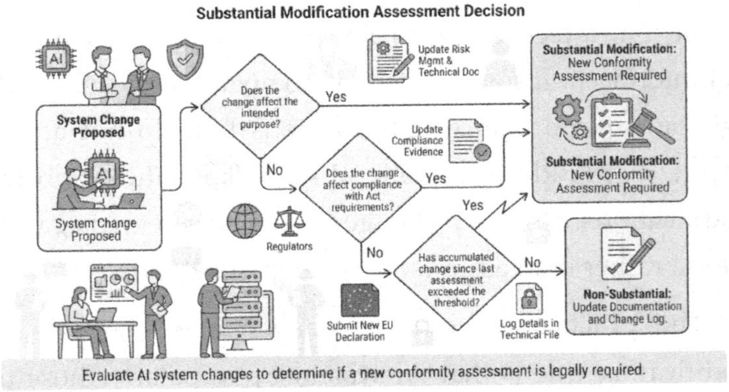

Building a Classification Methodology for the Enterprise AI Inventory

A classification methodology is the organizational process by which each AI system in the enterprise inventory is systematically evaluated against the Act's classification criteria and assigned a definitive risk category. Without a defined methodology, classifications are inconsistent across teams, subject to motivated reasoning (teams that want to avoid high-risk obligations may unconsciously underclassify), and difficult to defend in an audit.

A robust classification methodology includes several elements. First, a classification questionnaire that leads the assessor through the relevant criteria in a structured sequence: What is the system's intended purpose? Does that intended purpose fall within an Annex III domain? Is the system designed to be used as a safety component in a product covered by Union harmonization legislation listed in Annex I? The questionnaire should be specific enough to yield consistent answers across different assessors applying it to the same system.

Second, the methodology should include a governance checkpoint for borderline classifications. Many systems will fall clearly within or clearly outside the high-risk categories. But some will require interpretive judgment about whether their intended purpose falls within an Annex III domain or whether their use context qualifies them as safety components. Borderline classifications should be reviewed by the risk assessment lead and the legal function before being finalized, and the reasoning should be documented in detail, as borderline cases are most likely to attract regulatory attention.

Third, the methodology should include a periodic re-classification cycle. Classifications made at the time of initial system development may become stale as the system evolves, as the deployer's use of the system expands to new contexts, or as the regulatory definition of the Annex III categories is clarified through Commission guidance or enforcement decisions. An annual reclassification review for all systems in the inventory ensures that the organization's classification posture remains current.

The classification methodology should be documented in the organization's quality management system and subject to version control. When the methodology is updated, for example, to incorporate new regulatory guidance on a specific classification question, the update should be recorded, and the systems most likely to be affected by the updated guidance should be reviewed against the new version. Classification is a living determination, not a one-time judgment.

Diagram 6.3 - Enterprise AI Classification Methodology

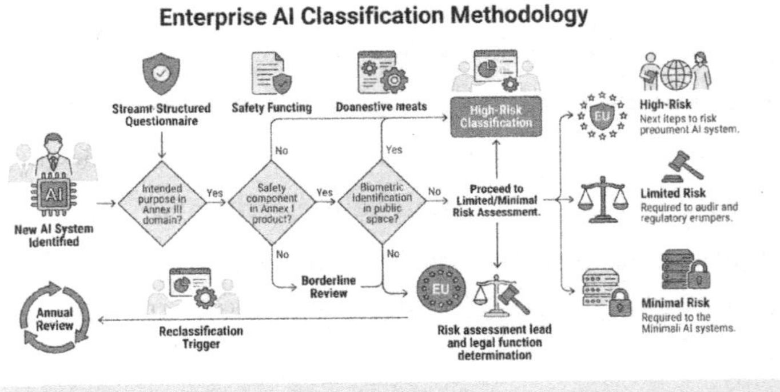

Systematic workflow for evaluating and managing regulatory risk across all enterprise AI deployments.

8.3 Core Obligations for High-Risk AI Providers

Articles 9 through 15 of the EU AI Act specify the substantive obligations that apply to providers of high-risk AI systems. These obligations are not a menu from which providers select the most convenient items. They have cumulative requirements that must all be satisfied before a high-risk AI system can be lawfully placed on the market or put into service. Understanding the substance of each obligation, and what 'meeting' actually means in an operational context, is the foundation of a compliant high-risk AI program.

Quality Management Systems: Structure and Requirements

Article 17 of the EU AI Act requires providers of high-risk AI systems to implement a quality management system that ensures compliance with the Act. The QMS must be documented, and its documentation must cover: the provider's strategy for regulatory compliance, including the techniques and processes used to ensure the AI system's compliance with the Act; the techniques and processes for AI system design; the examination, testing, and validation procedures; the technical specifications for the system; the supplier management systems; the conformity assessment procedures; the post-market monitoring plan; and the incident reporting mechanisms.

For organizations that already operate quality management systems under standards such as ISO 9001 or ISO 13485, applicable to medical devices, adapting an existing QMS to meet the Act's requirements is more efficient than building a new one. The key additions required by the Act are AI-specific: risk management for AI systems, training data governance, model testing and validation documentation, and post-market AI monitoring. Organizations without existing QMS infrastructure face a more substantial implementation effort but benefit from designing the QMS for AI compliance from the outset rather than retrofitting AI requirements into a system designed for conventional product quality.

The QMS must be applied consistently to every high-risk AI system in the provider's portfolio. It is not acceptable to have an organizational QMS that is not demonstrably applied to each specific system. The application of the QMS to a specific system is evidenced by the system-specific documentation generated by the QMS processes: risk management records, design review records, testing records, supplier assessment records, and post-market monitoring records. The QMS is the framework; the system-specific documentation is evidence that the framework was used.

Quality management for AI has characteristics that distinguish it from quality management for conventional software or hardware. AI system behavior is probabilistic, context-dependent, and affected by the quality and representativeness of training data in ways that conventional quality management frameworks were not designed to address. The QMS must include provisions for the specific quality challenges of AI: training data quality assurance, model validation across diverse operational subpopulations, performance monitoring under distributional shift, and the management of model updates and retraining cycles. Organizations adapting existing QMS frameworks should conduct a gap analysis specifically focused on AI-specific quality management requirements.

Diagram 6.4 - High-Risk AI Quality Management System Structure

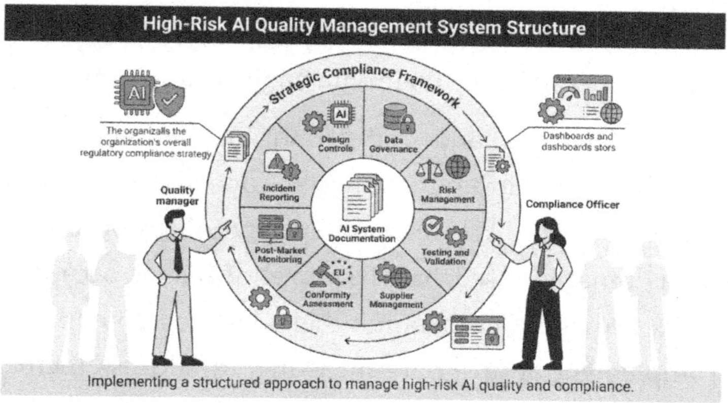

Data Governance and Training Data Documentation Standards

Article 10 of the EU AI Act imposes specific requirements on the data governance practices of providers of high-risk AI systems. Training, validation, and testing datasets must meet quality criteria: they must be relevant for the intended purpose, sufficiently representative, free of errors to the extent possible, and complete for the features and characteristics that matter for the system's design. The data governance requirements also cover the processes used to collect data, the examination of data for possible biases, and the measures taken to detect and address those biases.

The documentation requirement for training data goes beyond a simple description. The technical file must contain information on the origin of the data, on the scope and main characteristics of the datasets used, including their geographic coverage, the time period they cover, the language and demographic composition of the data subjects where applicable, and the data collection and labeling methodology. For many organizations, this level of documentation requires establishing data lineage tracking capabilities that have not historically been part of their data engineering infrastructure.

Bias detection and mitigation are among the most technically demanding aspects of data governance requirements. The Act does not specify which bias-detection methods must be used or what level of bias is acceptable; it requires providers to examine their data for potential biases and implement appropriate measures to address them. This formulation leaves significant interpretive latitude to providers but also places on them the burden of demonstrating that their bias examination was adequate, given the nature of the system and the potential harms of biased outputs. For a hiring algorithm, a cursory statistical review of aggregate output rates by demographic group is unlikely to satisfy the requirement; a more systematic analysis of disparate impact across multiple protected characteristics and intersectional subgroups is more consistent with the Act's purpose.

Special category personal data, such as health data, biometric data, and data revealing racial or ethnic origin, requires heightened attention in training data documentation because processing it for AI training may require specific legal bases under GDPR in addition to satisfying the Act's data quality requirements. The Act permits the use of special category data in training under specific conditions, including when technical limitations make it strictly necessary to ensure bias detection and correction, and when appropriate safeguards are in place. Organizations should document their legal basis for processing special category training data separately and ensure that their documentation covers both the

GDPR requirements and the Act's data governance requirements.

Diagram 6.5 - Training Data Governance Documentation Requirements

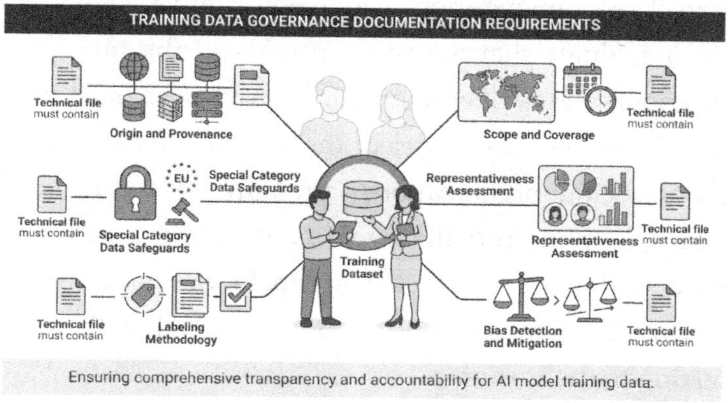

Human Oversight Design: Meaningful Control vs. Pro Forma Approval

Article 14 of the EU AI Act requires that high-risk AI systems be designed to allow natural persons to whom human oversight has been assigned to effectively oversee the system's functioning during its use. This requirement has a qualitative dimension that purely procedural compliance cannot satisfy: human oversight must be effective, not merely nominal. A requirement that a human 'review' an AI recommendation before it is acted on does not constitute meaningful oversight if the reviewer lacks the information, time, training, or authority to meaningfully evaluate and override the recommendation.

Meaningful human oversight has several identifiable characteristics. The human overseer must have access to the information necessary to evaluate the AI system's recommendation in context: the inputs the system used, the confidence or certainty indicators associated with the output, the system's performance characteristics in the relevant context, and the potential consequences of acting on an incorrect recommendation. The overseer must have sufficient time to evaluate that information without the time pressure that renders the review nominal. The overseer must have the authority to override or reject the system's recommendation without facing organizational pressure to conform to the system's output. And the overseer must have the technical understanding of the system's capabilities and limitations necessary to exercise genuine judgment rather than simply deferring to the system.

Organizations that have deployed AI systems with 'human in the loop' designs that do not meet these criteria must redesign the oversight mechanism before the system can be considered compliant. This is operationally consequential because redesigning human oversight often requires changes to user interface design, workflow design, staffing, and capacity planning (reviewers need enough time to review meaningfully), and training programs (reviewers need to understand what they are evaluating). These changes are not minor adjustments; they can require significant operational investment.

The technical documentation must describe the human oversight mechanism in specific operational terms: who performs the oversight function, what information they are presented with, how they exercise override authority, how override decisions are recorded, and how the oversight's effectiveness is monitored. Vague descriptions such as 'a human reviews all system recommendations' do not satisfy the documentation requirement because they do not demonstrate that oversight is effective. The description should be detailed enough for an auditor to verify that the documented oversight mechanism has been implemented in the actual deployment.

Diagram 6.6 - Meaningful Human Oversight Design Requirements

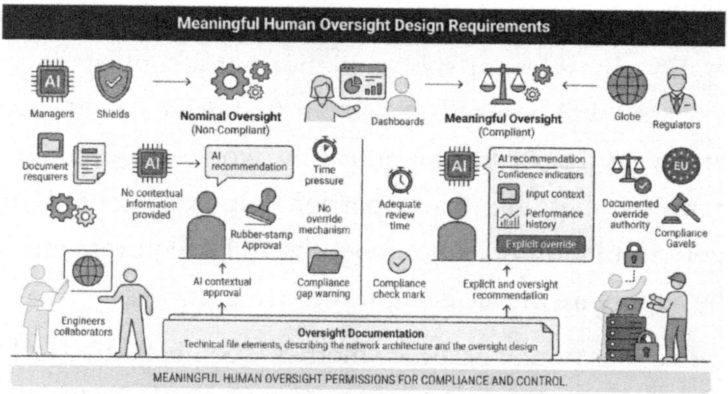

8.4 Accuracy, Robustness, and Cybersecurity Requirements

Article 15 of the EU AI Act requires that high-risk AI systems achieve an appropriate level of accuracy, robustness, and cybersecurity, and that they perform consistently in this respect throughout their lifecycle. This deceptively simple formulation encompasses a set of technical requirements that go significantly beyond conventional software quality standards and require AI-specific testing and validation methodologies.

Performance Metrics, Testing Obligations, and the Limits of Accuracy

The Act requires that the accuracy metrics and the relevant accuracy levels for high-risk AI systems be declared in the instructions for use. This requirement has a practical implication that many organizations underestimate: accuracy must be measured and declared not as a single aggregate metric but in a way that captures performance across the operationally relevant subpopulations and contexts in which the system will be used. A system that achieves 90% overall accuracy but performs significantly worse for demographic subgroups that are disproportionately represented in the target population does not meet the Act's requirements, even if the aggregate metric is high.

The selection of accuracy metrics is a substantive technical decision that must be documented and justified. For classification systems, the choice between precision, recall, F1 score, or other metrics depends on the asymmetry of error consequences in the specific application: a false positive in a credit denial system has different consequences than a false negative, and the metric selection should reflect that asymmetry. For regression systems used in risk scoring, the choice of error metrics and the specification of acceptable error bounds require domain expertise about what level of error is operationally significant in the application context.

Testing obligations extend beyond measuring accuracy on a held-out test set. The Act's requirement that systems 'perform consistently throughout their lifecycle' implies that testing must address performance not only in the training distribution but also under conditions that represent realistic operational variation. This includes testing data from different geographic regions or time periods than the training data, on edge cases and unusual inputs, and on data that simulates the kinds of degradation that may occur as the operational environment evolves. Robustness testing is the technical term for this class of evaluation, and it requires deliberate test design rather than simply measuring performance on a convenient test set.

The Act also requires that the instructions for use state the degree of accuracy that users can expect in specific operational contexts. This transparency obligation is operationally significant for deployers because it defines the conditions under which the system is warranted to perform at the declared level. Deployers who use the system outside those conditions cannot rely on the provider's conformity assessment to cover the system's performance in the novel context. This creates a practical obligation for deployers to verify that their operational deployment conditions match the conditions under which the system was assessed.

Diagram 6.7 - Multi-Dimensional Accuracy Assessment Framework

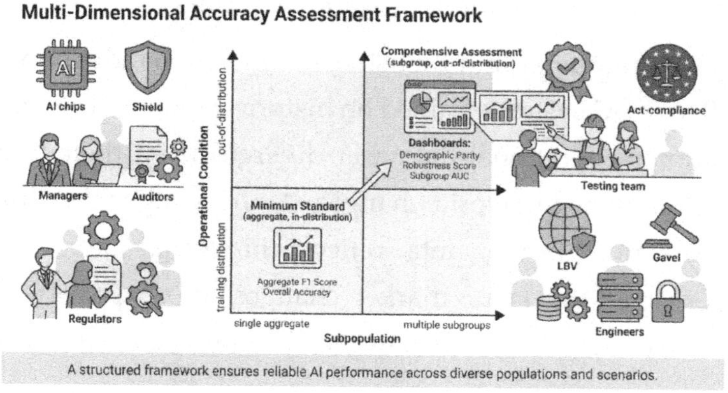

Robustness Against Adversarial Inputs and Distributional Shift

The Act's robustness requirement encompasses two distinct technical challenges. The first is resistance to adversarial inputs: inputs deliberately crafted to cause the AI system to produce incorrect or unexpected outputs. The second is performance stability under distributional shift: changes in the statistical characteristics of real-world inputs over time that degrade the system's performance without any deliberate adversarial action.

Adversarial robustness is particularly relevant for AI systems in high-risk domains because such systems may be targets of deliberate manipulation by individuals or organizations with an incentive to induce them to produce favorable outputs. A credit scoring system, a content moderation system, or a job application screening system may encounter applicants motivated to manipulate inputs to produce favorable determinations. The technical documentation must describe the measures taken to assess and address adversarial robustness, and the post-market monitoring plan should include mechanisms to detect patterns of adversarial behavior in operational inputs.

Distributional shift robustness addresses a more pervasive challenge: AI systems trained on historical data will eventually encounter operational data that diverges from their training distribution as the world changes. A hiring algorithm trained on historical hiring data reflects historical labor market patterns; as the labor market changes, those patterns may become less predictive or may encode outdated biases. A credit risk model trained on pre-pandemic data may not capture post-pandemic economic realities. The Act's requirement that systems 'perform consistently throughout their lifecycle' imposes an obligation to monitor distributional shifts and to assess their effect on system performance as part of the post-market monitoring program.

Testing for distributional shift robustness requires constructing test sets that simulate realistic future operational conditions, including data from time periods later than the training data, from geographic regions not well-represented in training, and from synthetic data designed to probe performance at the boundaries of the training distribution. These tests should be designed in collaboration with domain experts who understand how the operational environment is likely to evolve, rather than being generated purely algorithmically.

Diagram 6.8 - Robustness Testing: Adversarial and Distributional

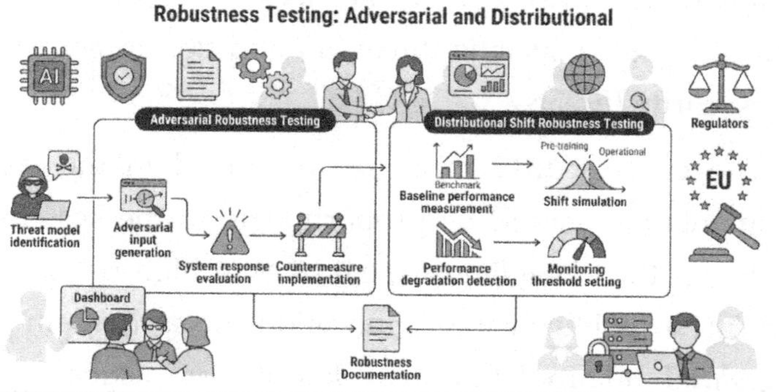

A two-track framework ensuring reliable system performance under attack or operational shifts.

Cybersecurity Requirements Specific to High-Risk AI Systems

The EU AI Act's cybersecurity requirements for high-risk AI systems are distinct from but complementary to the general cybersecurity requirements that apply to information systems under other EU legislation, including the NIS2 Directive. Article 15 requires that high-risk AI systems be resilient against attempts by unauthorized third parties to alter their use, outputs, or performance by exploiting system vulnerabilities. This encompasses both conventional cybersecurity threats and AI-specific attack vectors.

AI-specific attack vectors include model poisoning, where an attacker corrupts the training data or training process to embed vulnerabilities or biases into the model; model evasion, where an attacker crafts inputs that cause the model to produce incorrect outputs without triggering detection; and model extraction, where an attacker uses queries to the model to reconstruct confidential model parameters. The technical documentation must describe the measures taken to assess and address these AI-specific threats in addition to conventional cybersecurity threats.

The cybersecurity measures described in the technical documentation should be proportionate to the system's risk profile and the threat landscape it faces. A high-risk AI system used in a public-facing application in a domain where adversarial manipulation is a known risk, such as credit scoring or content moderation, requires more extensive adversarial robustness measures than a high-risk AI system used in an internal operational context with limited external access. The proportionality principle requires organizations to assess the realistic threat landscape for each system rather than applying a uniform set of cybersecurity measures across all high-risk AI systems.

Integration between the AI system's cybersecurity requirements and the organization's existing cybersecurity governance framework is essential for operational efficiency. The AI system's cybersecurity requirements should be documented in the technical file and then translated into specific security controls that are managed through the organization's existing security operations infrastructure: vulnerability management, patch management, access control, security monitoring, and incident response. The security operations team must be aware that the AI systems they are responsible for securing have specific AI-related vulnerabilities in addition to conventional software security vulnerabilities.

8.5 Post-Market Monitoring and Incident Reporting

The EU AI Act's post-market monitoring and incident reporting requirements complete the high-risk AI compliance lifecycle. They reflect the regulatory insight that conformity at the point of market placement is necessary but not a sufficient condition for safe and compliant AI systems: the real-world performance of AI systems must be continuously monitored, and when performance falls short of required standards or causes harm, that information must be captured, acted on, and reported to the appropriate authorities.

Designing a Post-Market Monitoring Plan

Article 72 of the EU AI Act requires providers of high-risk AI systems to implement a post-market monitoring plan that is proportionate to the nature of the AI technologies and the risks associated with the high-risk AI system. The plan must specify the methods and procedures to monitor system performance, the data collection mechanisms and data quality criteria for monitoring data, the performance metrics and thresholds to assess whether the system is performing as required, and the processes for reviewing monitoring data and acting on findings.

The monitoring plan should be developed in parallel with the technical documentation and conformity assessment, not as an afterthought added after the system is deployed. This is because the monitoring design depends on technical decisions made during system development: which logs the system generates, which monitoring hooks are built into the system architecture, and which performance measurement approaches are technically feasible given the system's design. Monitoring requirements that cannot be implemented without modifying the system are far more costly to address after deployment than during development.

Proportionality in monitoring plan design means that the intensity and sophistication of monitoring are calibrated to the system's risk profile. A high-risk AI system that makes recommendations affecting hundreds of thousands of individuals per day requires more intensive monitoring than one that is used in a specialized context affecting tens of individuals per month. The monitoring plan should specify the monitoring frequency, the statistical methods used to detect performance anomalies, the demographic subgroups for which monitoring data is collected and analyzed separately, and the thresholds that trigger a review or escalation.

The post-market monitoring plan must cover the entire deployment lifecycle, not just the initial deployment period. This means that the plan should specify how monitoring will be maintained and updated as the system evolves, how the monitoring data will be used to inform retraining and improvement decisions, and how monitoring findings will be incorporated into the annual update of the technical documentation. The plan is a living document that should be reviewed and updated as part of the system's regular governance cycle.

Deployers play an important role in monitoring post-market systems that they did not develop. The Act requires deployers of high-risk AI systems to monitor system performance against its intended purpose and to inform the provider without delay of serious incidents or malfunctions. This creates an obligation for deployers to implement their own monitoring infrastructure and establish a communication channel with the provider to report compliance-relevant findings. Organizations that deploy third-party high-risk AI systems should negotiate monitoring and reporting protocols into their supplier agreements at the procurement stage.

Diagram 6.9 - Post-Market Monitoring Plan Architecture

Post-Market Monitoring Plan Architecture

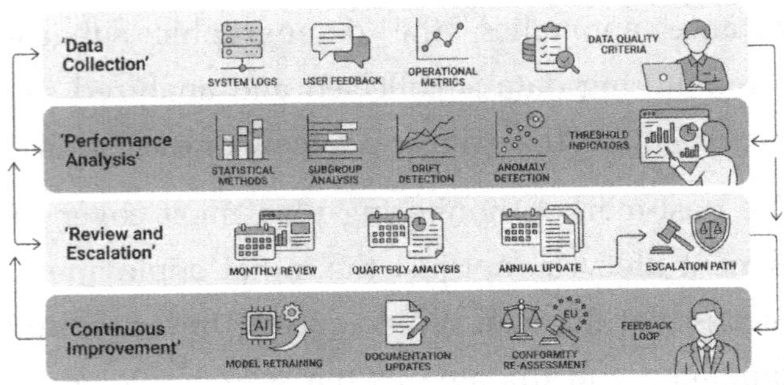

The pacin of monitoring findings provide one the short at businessen:gory leaurs.

Serious Incident Reporting: Thresholds, Timelines, and Notification

The EU AI Act defines a serious incident as a malfunction or performance deficiency of a high-risk AI system that directly or indirectly leads to the death or serious injury of a person, a serious and irreversible disruption of critical infrastructure, a serious breach of obligations under Union law intended to protect fundamental rights, or serious property damage or environmental harm. This definition sets a high threshold: not every performance issue is a serious incident requiring regulatory notification, but organizations must have processes capable of identifying when a performance issue crosses that threshold.

The notification timelines are tight and graduated. For incidents that cause the death of a person or involve serious injury, notification to the market surveillance authority must occur within fifteen days of becoming aware of the incident. For incidents involving an imminent risk of the death of a person or involving a serious, irreversible disruption of the provision of essential social services, notification must occur within three days. For all other serious incidents, notification must be made within 15 days. These timelines begin when the provider or deployer becomes aware of the incident, not when the incident occurred, which creates a premium on rapid incident detection and triage.

The content of the incident notification must cover the identity of the provider and the AI system, the nature and severity of the incident, the harm caused or risk of harm, the suspected cause of the incident, and the corrective measures taken or planned. Organizations should prepare notification templates and a drafting process that can be activated quickly in the event of a serious incident, because assembling coherent and accurate notification content under time pressure without preparation is extremely difficult. The notification templates should be maintained in the incident management runbook.

The serious incident reporting obligation applies to both providers and deployers, but the obligations differ between them. Providers must report to the market surveillance authority of the member state in which the incident occurred or in which the system is placed on the market. Deployers must report to providers and, if they become aware of a serious incident before the provider has been notified, directly to the market surveillance authority. This dual reporting structure means that organizations must know their role in any specific incident scenario and have clear internal processes for determining whether direct supervisory notification is required.

After notification, providers and deployers are expected to cooperate with market surveillance authorities in investigating the incident, providing additional documentation as requested, and implementing corrective measures. The Act allows market surveillance authorities to require that AI systems be withdrawn from the market or that their operation be suspended if the investigation into the incident reveals that the system poses an unacceptable risk. Cooperation with the investigation and prompt implementation of corrective measures is both a legal obligation and the practical path to the shortest possible enforcement exposure.

Diagram 6.10 - Serious Incident Reporting: Thresholds and Timelines

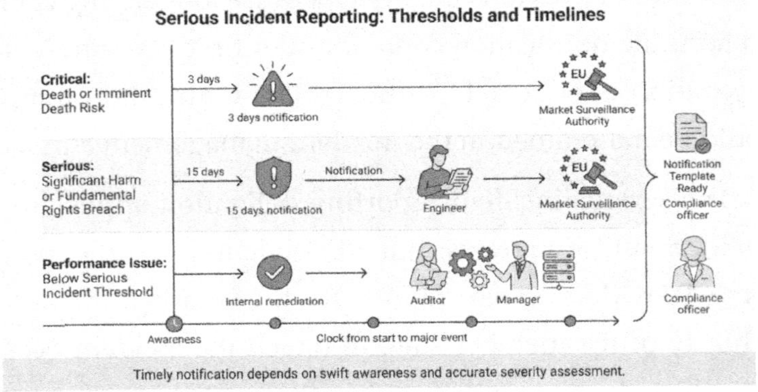

Integrating Monitoring Data into Quality Management and System Improvement

The regulatory requirement to conduct post-market monitoring is also an operational opportunity. The monitoring data generated by deployed high-risk AI systems is among the most valuable data the organization can collect about those systems' real-world performance. It reveals how the system behaves across the full diversity of real-world inputs, which are always more varied than the test set. It identifies edge cases and failure modes that were not anticipated during development. It tracks the evolution of system performance over time as the operational environment changes. Organizations that treat this data purely as a compliance artifact miss significant opportunities to improve their AI systems and their understanding of the domains in which those systems operate.

Integrating monitoring data into the quality management system means establishing formal processes for reviewing monitoring findings, assessing their implications for system quality and conformity, and translating those assessments into development priorities, retraining decisions, or documentation updates. The review process should be regular, structured, and attended by the technical and compliance stakeholders who have both the context to interpret the findings and the authority to act on them. A monitoring dashboard that is observed but not acted on represents the technical infrastructure of monitoring without its operational substance.

The feedback loop between post-market monitoring and system improvement should be documented in the quality management system. When monitoring findings identify a performance issue that warrants corrective action, the corrective action should be tracked through the QMS as a formally documented improvement activity: the finding is documented, the root cause is analyzed, the corrective action is specified and implemented, and the effectiveness of the corrective action is verified through subsequent monitoring. This closed-loop process is the mechanism by which post-market monitoring contributes to the continuous improvement of system quality rather than merely documenting its current state.

For providers of systems used by multiple deployers across different operational contexts, aggregating monitoring data across deployers provides insights that no individual deployer's monitoring can reveal, patterns that span deployment contexts, systematic performance issues in specific input domains, or emerging trends that are not yet visible in any individual deployment. Providers should design their post-market monitoring infrastructure to aggregate and analyze cross-deployer monitoring data, with appropriate data governance arrangements that protect deployer confidentiality while enabling the systemic analysis required by the Act's monitoring purpose.

8.6 Manager's Checklist

Use this checklist to evaluate your organization's high-risk AI compliance posture. Each item addresses a specific obligation or operational practice required for compliant high-risk AI management. The checklist applies to both providers and deployers, with noted distinctions where the obligations differ.

- Every AI system in the organization's portfolio has a documented risk classification determination using a defined methodology, with borderline cases reviewed by the risk assessment lead and legal function.
- A documented substantial modification determination process exists and is applied to every system change, with a change accumulation tracking mechanism that monitors cumulative modifications since the last conformity assessment.

- The quality management system covers AI-specific quality requirements, including training data governance, model testing and validation, and AI-specific performance monitoring. It is demonstrably applied to each high-risk AI system through system-specific documentation.
- Training data documentation for each high-risk AI system covers origin, scope, representativeness, bias examination methodology and findings, mitigation measures, and is stored as part of the technical file.
- The human oversight mechanism for each deployed high-risk AI system is designed for meaningful oversight: reviewers have access to contextual information, adequate review time, documented override authority, and training on the system's capabilities and limitations.
- Accuracy metrics for each high-risk AI system are measured and declared by subpopulation and operational context, not only as aggregate metrics, and the instructions for use specify the operational conditions under which the declared accuracy applies.
- A post-market monitoring plan exists for each high-risk AI system, is implemented with the specified monitoring infrastructure, and includes subgroup performance monitoring, drift detection, and a defined review and escalation process.
- Incident detection is configured with compliance-relevant thresholds for each deployed high-risk AI system, and incident response runbooks, including

notification templates for serious incident reporting, have been prepared and have been tested.

- For systems procured from external providers, deployer obligations, including monitoring implementation, log maintenance, and provider notification processes, are assigned to named internal owners with documented procedures.
- For deployers subject to fundamental rights impact assessment requirements, the assessment has been conducted before deployment, is documented, and is reviewed periodically or when the system's use context changes significantly.

8.7 Takeaway

High-risk AI management is not about achieving a compliance certificate and moving on. It is about building and sustaining the operational infrastructure that allows an organization to deploy AI in consequential domains responsibly: with documented evidence of what the system does and under what conditions, with human oversight that is designed to be genuinely effective, with monitoring that detects real performance issues before they become serious incidents, and with the organizational readiness to respond to incidents transparently and promptly. The Act's requirements for high-risk AI systems are demanding because the consequences of getting them wrong are serious, for the individuals affected by AI decisions and for the organizations that make them.

The provider-deployer framework is one of the Act's most important structural contributions to AI governance because it distributes responsibility proportionally to capability: providers who design and build systems bear the primary obligations for what those systems are capable of doing. In contrast, deployers who choose to use those systems in specific operational contexts bear the obligations that arise from their actual deployment. Managers on both sides of that relationship need to understand their obligations specifically, build the infrastructure to meet them, and engage in the ongoing dialogue between providers and deployers that the Act envisions as the mechanism for keeping high-risk AI systems aligned with their conformity status throughout their operational lives.

9 The Foundation Model Problem

A chief product officer at a mid-sized European enterprise software company receives a legal memo in March of the Act's application year. The memo has one central question: Does the company's flagship product, a contract analysis tool built on top of a major foundation model API, make the company a provider of a general-purpose AI system, a deployer of a high-risk AI system, or both? The product team has always described itself as a software company, not an AI company. The legal team is no longer confident that distinction matters. The company does not train its own model. It calls an API, passes user-uploaded contract documents through a prompt template designed by the engineers, and returns a structured analysis. But that prompt template constitutes a non-trivial design choice, and recent guidance suggests that sufficiently substantial modifications to a foundation model's behavior may shift regulatory responsibility toward the integrating organization.

The chief product officer schedules a working session with legal, engineering, and procurement. Before the meeting begins, no one in the room can agree on the basic facts: which foundation model version is running in production, what the API contract says about liability and compliance obligations, whether the model provider has published the required technical documentation, and what changes to the prompt architecture over the past eighteen months might constitute a modification under the Act. That situation, replicated across thousands of enterprises that have built on foundation model APIs without tracking the compliance implications, is the foundation model problem in concrete form.

9.1 Why It Matters

The EU AI Act's treatment of general-purpose AI models represents one of the most consequential and least-understood elements of the regulatory framework for most enterprise managers. The obligations imposed on GPAI model providers are extensive, but they sit at a layer of the AI supply chain that most enterprises do not occupy. What matters for the majority of organizations is the downstream effect: the compliance obligations that flow from a decision to build a product or service on top of a foundation model API, the contractual protections that must be secured from the model provider, and the conditions under which a downstream integrator becomes, in the Act's terms, a provider in its own right.

Understanding the GPAI framework is not optional for managers whose organizations use commercially available large language models, image generation models, or multimodal AI systems in their products or operations. The framework determines what documentation you are entitled to receive from your model vendor, what disclosures you may need to pass on to your own downstream users, and what happens to your compliance posture if the foundation model you depend on is redesignated as a systemic-risk model. These are operational questions with legal consequences, and they require answers before a product is deployed, not after a regulatory inquiry has begun.

The foundation model problem is also a vendor management problem. As foundation models become the infrastructure for enterprise AI, the relationship between the enterprise and its model provider takes on characteristics of a regulated supply chain rather than an ordinary software subscription. The Act creates obligations that flow upstream and downstream, and organizations that understand those flows will be better positioned to negotiate contracts, conduct due diligence, manage liability, and maintain compliance as the model provider's own regulatory status evolves.

9.2 Defining General-Purpose AI Under the Act

The Act's definition of general-purpose AI is intentionally broad, which creates classification questions that every organization using large-scale AI models must answer. Before addressing compliance obligations, a manager must be clear on what qualifies as a GPAI model, how GPAI models differ from GPAI systems, and how the GPAI framework interacts with the separate high-risk classification that governs many enterprise AI use cases.

The GPAI Definition: What Qualifies and What Does Not

The Act defines a general-purpose AI model as an AI model trained on large amounts of data using self-supervision at scale, that displays significant generality, and that is capable of performing a wide range of distinct tasks. The key phrase is 'wide range of distinct tasks.' A model that is trained and fine-tuned exclusively for a single narrow function, such as classifying images of manufacturing defects into predefined categories, does not qualify as a GPAI model under this definition. A model that can write prose, summarize legal documents, answer questions in multiple languages, generate code, and perform sentiment analysis qualifies, because its competence spans domains in a way that was not task-specifically engineered.

The significance of the GPAI designation is that it triggers a separate tier of obligations on the model's developer. These obligations apply regardless of whether the model is ultimately used in a high-risk application. A large language model used only for drafting internal HR communications is still a GPAI model if it meets the generality threshold. The GPAI framework thus adds an orthogonal compliance dimension: a model can be both a GPAI model and a component in a high-risk AI system, and each classification produces its own set of requirements. Managers need to understand both layers, not just the one that appears most salient from their product's use case.

The Act explicitly excludes the GPAI definition models released under a free and open-source license, subject to important conditions. Also, it excludes models used exclusively for research and development purposes before any market placement. These carveouts are discussed in detail in the final section of this chapter. Still, it is worth noting at the outset that the open-source exception is narrower than many enterprises assume: it applies to the model provider's obligations, not necessarily to the downstream integrator's, and it does not extend to systemic-risk models regardless of licensing.

Diagram 7.1 - GPAI Model Classification Decision Tree

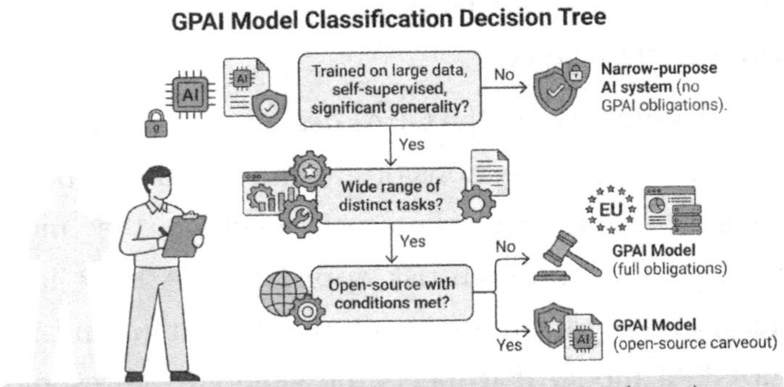

Guidelines for classifying General Purpose AI models under current regulatory frameworks.

The Distinction Between GPAI Models and GPAI Systems Built on Those Models

One of the most practically important distinctions in the Act is the difference between a GPAI model and a GPAI system. A GPAI model is the trained model itself: the weights, architecture, and associated technical artifacts. A GPAI system is an AI system based on a GPAI model that can serve various purposes. The company that trains GPT-class models is a GPAI model provider. The enterprise that wraps those model weights in an application configures prompts, adds retrieval pipelines, sets behavioral constraints, and deploys the result to end users is operating a GPAI system and, potentially, a high-risk AI system, depending on the use case.

This distinction matters because the obligations differ. GPAI model providers must maintain technical documentation, publish model cards, establish copyright compliance policies, and, if their model crosses the systemic-risk threshold, conduct adversarial testing and report incidents to the AI Office. GPAI system operators, by contrast, face obligations that depend on the system's use rather than on the properties of the underlying model. If the system is used in a high-risk context, the operator faces the full high-risk obligations. If the system is used in a limited-risk context, the transparency obligations in Articles 50 through 52 apply. If the system is minimal risk, there are no hard legal requirements beyond general product liability.

The layered structure creates a compliance analysis problem that many enterprises have not yet solved. When an enterprise deploys a chatbot built on a foundation model API, it must first classify the underlying model (GPAI or not), then classify its own system (high-risk, limited-risk, or minimal-risk based on the use case), and then determine which obligations it inherits from the model provider's compliance posture and which it must independently satisfy. Maintaining a clear map of this layered classification for every AI system in production is a foundational task for any compliance function.

Diagram 7.2 - GPAI Model vs. System Layers

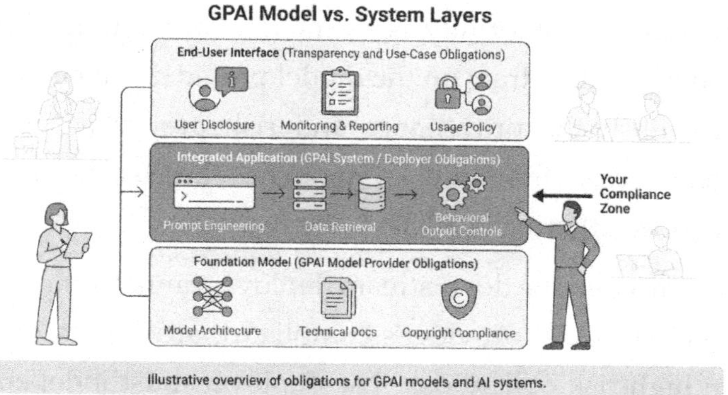

Illustrative overview of obligations for GPAI models and AI systems.

How the GPAI Framework Interacts with High-Risk Classification

An AI system built on a GPAI model that is deployed in a high-risk use case must satisfy both frameworks simultaneously. The high-risk framework, detailed in Annex III of the Act, governs the system's functionality in specific application domains, such as employment decisions, credit scoring, biometric identification, and critical infrastructure management. The GPAI framework governs the training, documentation, and systemic impact of the underlying model. An enterprise that deploys a foundation-model-powered screening tool for job applications is subject to the high-risk framework for the tool's behavior and the GPAI framework for the model it runs on.

The interaction between these frameworks creates both compliance obligations and compliance resources. A GPAI model provider that has published complete technical documentation, established a copyright compliance policy, and made a machine-readable training data summary available has already produced artifacts that a downstream deployer can incorporate into its own conformity assessment. The high-risk conformity assessment required under Articles 9 through 14 may draw on the model provider's documentation to establish the provenance and risk properties of the AI system's core inference component, thereby reducing the evidence-gathering burden on the deployer.

However, the downstream deployer cannot simply rely on the GPAI model provider's compliance artifacts to satisfy its own high-risk obligations. The deployer must independently demonstrate that the system, as deployed, including all integrations, prompt configurations, data pipelines, and output handling logic, satisfies the high-risk requirements. The model provider's documentation is an input to that analysis, not a substitute for it. Managers who assume that using a compliant foundation model automatically makes their application compliant are taking a position that regulators are unlikely to accept.

Diagram 7.3 - Dual Framework Compliance Overlap

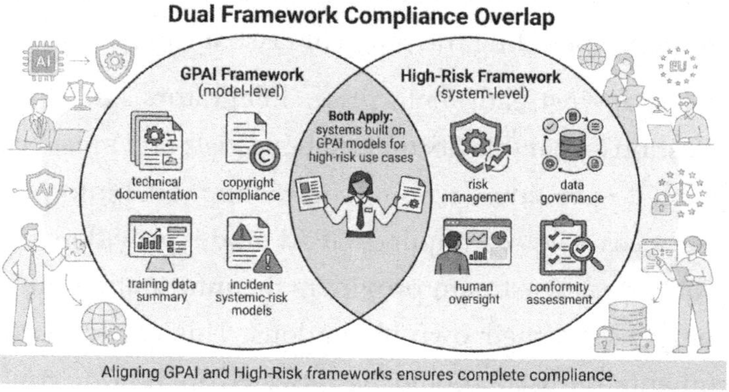

9.3 Transparency Obligations for GPAI Providers

For enterprises that procure foundation model capabilities from external vendors, understanding the Act's transparency obligations for those vendors is essential. These obligations define what information you are legally entitled to receive, what you can reasonably demand in a vendor contract, and what you can incorporate into your own downstream compliance documentation. They also define the floor of what responsible AI supply chain management looks like under the Act.

Technical Documentation Requirements for GPAI Models

Article 53 of the Act requires GPAI model providers to maintain technical documentation sufficient to enable competent authorities to assess compliance with the Act's requirements. The content of that documentation is specified in Annex XI. It includes information about the model's training methodology, the nature and sources of training data, the computational resources used, energy consumption, performance benchmarks across standard evaluation tasks, known limitations and failure modes, and the model's capabilities, including the range of tasks it can perform.

For downstream enterprise integrators, the practical implication is that this documentation exists and must be made available to providers who integrate the model into products and services. The Act requires GPAI model providers to make available to downstream providers the information they need to comply with their own obligations. This creates a right to documentation that should be reflected in vendor contracts. An enterprise that signs a foundation model API agreement without securing explicit provisions for access to Annex XI-compliant technical documentation is operating without information it is legally entitled to receive.

In practice, major foundation model providers have begun publishing model cards, system cards, and technical reports that approximate the requirements of Annex XI. However, the quality and completeness of these disclosures vary considerably, and the information in a public model card may not match that in a compliance-grade technical documentation package. Enterprises should distinguish between public marketing documentation and the technical documentation required by the Act to be made available for compliance purposes. They should request the latter explicitly when establishing vendor relationships.

The documentation requirement also has a temporal dimension. Technical documentation must be updated when the model changes in ways that affect its capabilities, limitations, or risk profile. An enterprise that relied on a model provider's documentation at the time of contract signing needs a mechanism to receive updated documentation when the model is retrained, fine-tuned, or its configuration changes materially. Version tracking for foundation model documentation is an underappreciated compliance obligation that only becomes visible when a downstream enterprise is asked to demonstrate that its conformity assessment is current.

Diagram 7.4 - Annex XI Documentation Components

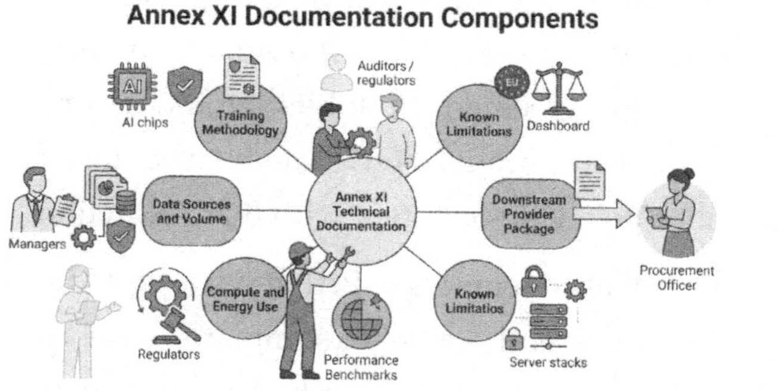

Ensuring comprehensive technical transparency for Annex XI compliance.

Copyright Compliance Policies and Training Data Disclosure

Article 53 also requires GPAI model providers to establish and make publicly available a policy for complying with Union copyright law, particularly with respect to training data. The requirement stems from recognized tension between large-scale AI training practices and intellectual property rights under the EU Copyright Directive, which includes a text-and-data-mining exception subject to rights-holder opt-out provisions. A GPAI provider must demonstrate that it has a policy for respecting those opt-outs and for addressing copyright claims related to its training data.

For downstream enterprise deployers, this matters in at least two ways. First, if your organization's products use output from a foundation model, and a third party asserts that the model was trained on their copyrighted content in violation of their opt-out rights, the copyright liability question may extend back through the supply chain. Understanding the model provider's copyright compliance policy is therefore part of your intellectual property risk management, not just your AI Act compliance. Second, if your own organization has content that you do not want used for AI training, you need to ensure that your API agreement prohibits the model provider from training on your inputs and that the provider's copyright compliance policy is consistent with that restriction.

The Act also requires GPAI model providers to make available a sufficiently detailed summary of the training data used to train the model, described in a way that enables downstream providers and deployers to understand the nature of the training corpus without necessarily disclosing proprietary competitive information. This summary is a required compliance artifact, and downstream enterprises should verify that their model provider has published one. The absence of a compliant training data summary is a potential red flag in due diligence and should be included in any vendor compliance assessment.

Diagram 7.5 - Copyright Compliance Policy Chain

Machine-Readable Summaries and Downstream Deployer Information Requirements

The Act anticipates that compliance information will need to flow through multi-tier supply chains in machine-readable formats that can be processed by automated systems and integrated into compliance tooling stacks. GPAI model providers are required to provide information to downstream providers in a form that enables those providers to comply with their own transparency and documentation obligations under the Act. In practice, this means that the compliance artifacts produced by a foundation model provider, including technical documentation, training data summaries, and copyright compliance policy statements, should be available in structured, versioned, machine-readable formats.

For enterprise compliance functions, this creates an opportunity to automate parts of the vendor compliance monitoring process. If a model provider's documentation is available in a structured format, it can be ingested into a compliance tooling stack, compared against policy requirements, flagged when it changes, and incorporated into conformity assessment documentation without requiring manual extraction and reformatting. Organizations that build their compliance workflows to accept structured vendor compliance data will have a material operational advantage over those that rely on PDF review cycles.

The downstream deployer information requirement also creates an obligation that flows in the other direction: from the enterprise integrator to any organization that builds on top of its platform. If your organization provides AI-enabled services to other businesses, and those businesses use your services to build their own products, you become a link in the information chain that the Act is trying to make transparent from end to end. The information you receive from your model provider must be reconciled with the information you are required to pass downstream to your own customers. Tracking this obligation requires awareness of your position in the AI supply chain and deliberate planning for information flow management.

Diagram 7.6 - Machine-Readable Compliance Data Flow

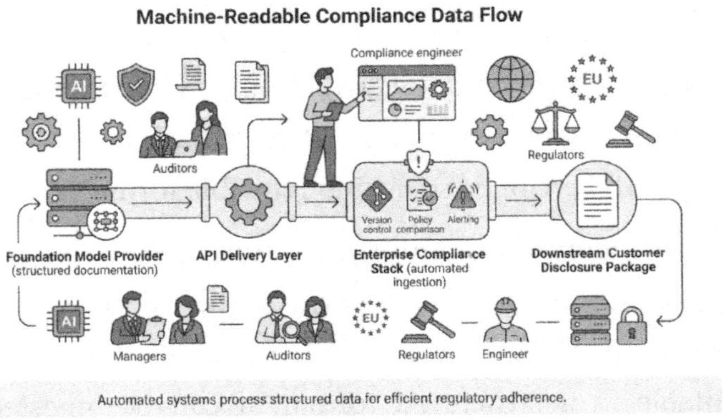

9.4 Systemic Risk: The Top Tier of GPAI Regulation

The Act creates a two-tier framework within the GPAI category: standard GPAI models and GPAI models with systemic risk. The systemic-risk designation is reserved for models whose scale and capabilities pose risks that extend beyond individual deployments, affecting entire sectors, infrastructure systems, or democratic institutions. For most enterprise compliance managers, the systemic-risk tier matters not because their organization is developing such models, but because they are likely using them.

How Systemic Risk Is Defined and What Triggers Designation

The Act defines systemic risk for GPAI models primarily by reference to a computational threshold: models trained using more than 10^25 floating-point operations (FLOPs) are presumed to pose systemic risk. This threshold is derived from research suggesting that models trained at this scale tend to exhibit emergent capabilities, meaning behaviors that were not specifically designed or anticipated, and that these emergent capabilities increase the potential for unexpected and difficult-to-control downstream effects. The 10^25 FLOPs threshold is not the only basis for designation: the AI Office may also designate a model as systemic-risk based on its actual capabilities, market reach, or the nature of the risks it poses, even if it was trained below the computational threshold.

The systemic-risk designation is significant because it triggers a substantially more demanding set of obligations on the model provider, including adversarial testing, incident reporting to the AI Office, enhanced cybersecurity measures, and cooperation with the AI Office's supervisory activities. It also signals to downstream deployers that the model they are using has been assessed as posing elevated societal risk, with implications for their own risk management obligations. Deployers building on a systemic-risk model face a different risk calculus than those building on a standard GPAI model. That difference should be reflected in their conformity assessments and post-market monitoring plans.

The designation process can be initiated by the AI Office, by a model provider's own assessment (providers must notify the AI Office if they believe their model meets the threshold), or by a competent national authority bringing information to the AI Office's attention. Designation is not permanent: a model that is subsequently modified to reduce its capabilities, or that the AI Office determines no longer poses systemic risk, may have its designation revised. Downstream deployers need monitoring mechanisms to track their primary foundation model provider's designation status and to trigger internal compliance reviews when that status changes.

Diagram 7.7 - Systemic Risk Designation Pathway

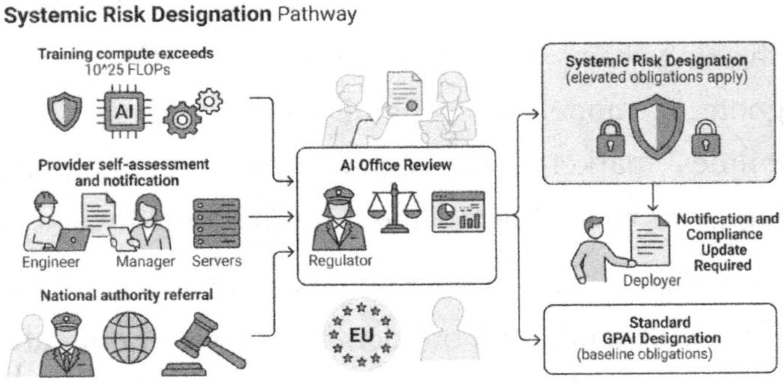

Illustrative process for determining systemic risk in general-purpose AI systems.

Adversarial Testing, Red-Teaming, and Model Evaluation Requirements

Systemic-risk model providers are required to conduct model evaluations and adversarial testing, commonly referred to as red-teaming, to identify and mitigate the risks that their models may pose across a range of potential deployments. The Act does not prescribe a specific red-teaming methodology. Still, it requires testing before a model is placed on the market and ongoing testing as the model evolves. The AI Office is expected to develop and publish guidelines and methodologies for this testing, and harmonized standards bodies are working on evaluation frameworks that will provide more specific technical guidance.

The significance of adversarial testing requirements for downstream enterprise deployers is twofold. First, the existence of adversarial testing results is information that downstream deployers should seek access to in their vendor due diligence. If your model provider has conducted red-teaming exercises on the model you are using, the findings of those exercises are directly relevant to your own risk assessment. A contract that does not provide you with access to red-team summaries for systemic-risk models leaves a material gap in your compliance posture. Second, the adversarial testing obligation does not substitute for the deployer's own security and safety testing. The model provider's red-team results cover the model's general capabilities; your organization must separately assess how those capabilities interact with your specific deployment context, data inputs, and user population.

The model evaluation requirement also has implications for enterprise procurement processes. Organizations selecting foundation model providers should include the availability and completeness of model evaluation results as a formal evaluation criterion. The question 'Has this model been red-teamed, and can you share a summary of findings?' should appear in every RFP for foundation model services. The absence of a credible answer is a selection-disqualifying finding for organizations that are deploying in high-risk or sensitive contexts.

Diagram 7.8 - Red-Teaming Obligation and Downstream Use

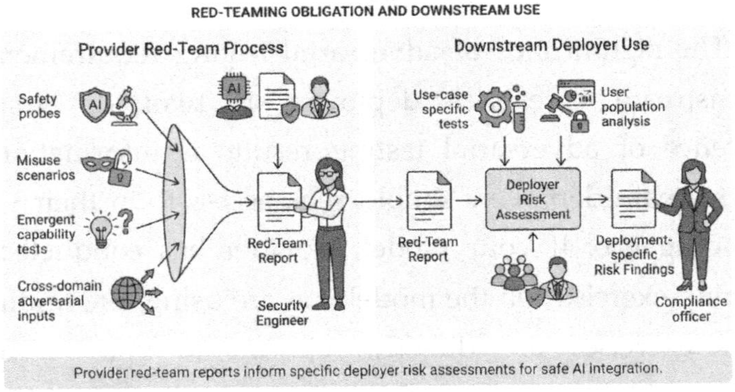

Incident Reporting Obligations and Cooperation with the AI Office

Systemic-risk GPAI model providers are required to report serious incidents to the AI Office. Under the Act, a serious incident in the GPAI context includes incidents in which the model's behavior caused or contributed to a death or serious harm to a natural person, a significant disruption of critical infrastructure, or a risk to public security. The reporting obligation applies regardless of whether the incident originated from the provider's own deployment of the model or from a downstream deployer's use of it, creating an information flow obligation between downstream deployers and their model providers.

For enterprise deployers, this creates a contractual and operational need: you must be able to detect, assess, and report to your model provider any serious incident that involves the model's behavior in your deployment environment, and your contract must specify how that reporting process works. A deployer who experiences a serious incident involving a systemic-risk model and fails to report it to the provider may impede the provider's ability to fulfill its regulatory obligations to the AI Office, thereby exposing the deployer to its own regulatory liability.

The cooperation obligation is broader than incident reporting. Systemic-risk model providers are required to cooperate with the AI Office in any investigation it undertakes, including by providing access to model weights, documentation, and evaluation results upon request. While this cooperation obligation falls on the provider rather than the downstream deployer, it creates a transparency expectation that, in turn, indirectly affects the enterprise: the AI Office may, in the course of an investigation, develop findings about the model's capabilities or risks that affect the deployer's own compliance position. Monitoring for AI Office guidance and investigation outcomes related to the foundation models you use is a component of operational intelligence for any enterprise compliance function.

Diagram 7.9 - Incident Reporting Chain for Systemic-Risk Models

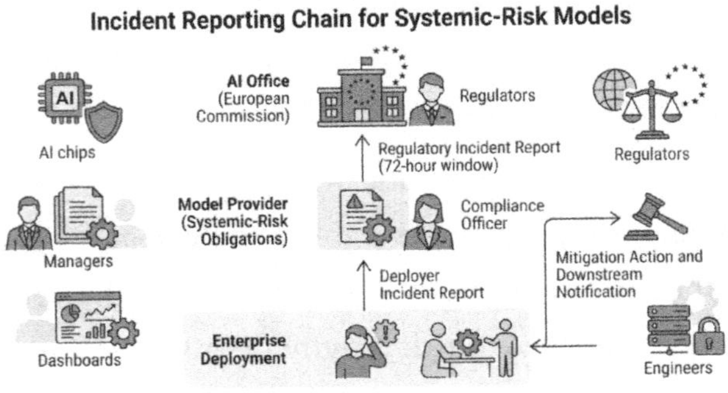

Standardized incident reporting flow for high-impact AI systems.

9.5 Downstream Liability and the Deployer's Position

The most operationally complex aspect of the GPAI framework for most enterprise managers is not what their model provider must do, but what they themselves must do as organizations that build on foundation model capabilities. The Act establishes a nuanced allocation of responsibility between GPAI model providers and downstream deployers. That allocation is shaped by the contracts they sign, the modifications they make, and the use cases they enable. Getting this allocation right is the foundation of a defensible compliance posture for any enterprise integrator.

How Liability Is Allocated Between GPAI Providers and Downstream Deployers

The Act's allocation of responsibility between GPAI model providers and downstream deployers follows a general principle. Each party in the AI supply chain is responsible for the obligations that fall within its sphere of control. The GPAI model provider is responsible for the model's training, documentation, copyright compliance, and, if applicable, adversarial testing and incident reporting. The downstream deployer is responsible for configuring, integrating, and deploying the model for a specific use case, and for ensuring that the resulting AI system complies with the obligations applicable to that use case.

This allocation can shift in important ways. If a downstream deployer makes substantial modifications to a GPAI model, they may take on provider-level obligations for the modified model. The Act also provides that where a downstream deployer places an AI system on the market under its own name, that deployer is treated as a provider for regulatory purposes, regardless of whether it trained the underlying model. These provisions are designed to prevent a simple division of labor in which the nominal model developer claims compliance responsibility and the actual product deployer disclaims it, leaving no party accountable for the system that end users actually interact with.

The practical implication is that enterprise compliance teams must carefully analyze their deployment architecture for any activity that could be characterized as a substantial modification or market placement under their own name. Adding a system prompt that materially changes the model's behavior, training the model on proprietary data through fine-tuning, building a user interface that presents the AI capability as the enterprise's own product, or selling access to the configured system as a service may each, depending on the specifics, push the deployer toward provider status. These are legal determinations that require early involvement of legal counsel and a clear factual record of what changes were made to the base model.

Diagram 7.10 - Liability Allocation: Provider vs. Deployer

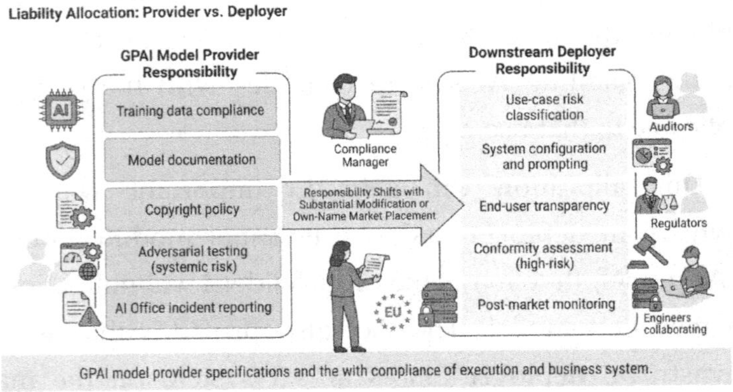

Contractual Protections and Due Diligence When Building on Foundation Models

For enterprises that integrate foundation model APIs into their products and services, the vendor contract is the primary mechanism for managing GPAI compliance risk. A well-drafted foundation model service agreement should address several specific compliance topics: the provider's obligation to supply and maintain Annex XI-compliant technical documentation; notification procedures when the model is updated, retrained, or its designation status changes; representation and warranty regarding copyright compliance policy; access to red-team findings for systemic-risk models; the definition of 'permitted use' in a way that explicitly covers the enterprise's intended applications; indemnification provisions addressing third-party IP claims arising from training data; and data retention and processing provisions consistent with the enterprise's obligations under GDPR and the Act.

Due diligence before signing a foundation model agreement should be treated as a compliance-grade review, not a standard software procurement review. The due diligence process should include a technical assessment of the model's documentation against Annex XI requirements, a review of the provider's published copyright compliance policy for completeness and enforceability, a request for the training data summary, an assessment of whether it is sufficient to support downstream deployer compliance needs, and a legal review of the contract's liability allocation provisions against the Act's requirements. For systemic-risk models, due diligence should also include a request for red-team findings and an assessment of the provider's incident-reporting procedures.

Post-contract management is equally important. A foundation model contract that was fully compliant at signing may become inadequate if the model provider changes its documentation practices, updates the model in ways that affect the deployment's compliance posture, or receives a systemic-risk designation after the contract was signed. Compliance teams should establish a schedule for periodic vendor compliance reviews, define the triggers that initiate out-of-cycle reviews (including model version changes and designation status changes), and maintain a file of communications with the model provider on compliance topics to document the enterprise's diligence over time.

Open-Source GPAI Carveouts and Their Limits

The Act provides a partial carveout for GPAI models released under a free and open-source license. Providers of open-source GPAI models are exempt from the technical documentation and copyright compliance policy requirements, on the theory that the openness of the model weights, training code, and associated artifacts provides a form of transparency that compensates for the absence of formal documentation obligations. This carveout has significant implications for enterprises that use open-source foundation models rather than commercial API services, because the compliance information that would otherwise flow from provider to deployer through contractual and regulatory mechanisms is instead embedded in the model's public release artifacts.

The open-source carveout has three important limits that enterprise managers must understand. First, the carveout does not apply to open-source GPAI models that the AI Office has determined to pose systemic risk. An open-source model trained above the 10^25 FLOPs threshold, or designated as systemic-risk on capabilities grounds, is subject to the full systemic-risk obligations regardless of its licensing terms. Second, the carveout applies to the model provider's obligations, not to the downstream deployer's. An enterprise that fine-tunes, configures, and deploys an open-source foundation model must still satisfy all the obligations that apply to its own deployment, including high-risk conformity assessment if the use case warrants it. Third, the carveout applies only when the model weights, training methodology, and associated data are made genuinely available to the public, not when a model is described as open-source but provided only through a commercial API with restricted access to the underlying artifacts.

For enterprises choosing between commercial and open-source foundation models, the compliance implications of the open-source carveout are not uniformly in favor of open-source. A commercial provider with robust documentation practices and a strong commitment to contractual compliance may offer a more tractable compliance path than an open-source model with incomplete documentation and unclear provenance of training data. The choice between commercial and open-source foundation models should explicitly include a comparative compliance analysis, not just a cost and capability analysis. The pilot-to-production pathway for open-source AI is often longer, not shorter, when compliance requirements are properly accounted for.

When a Deployer Becomes a Provider Through Substantial Modification

The Act's treatment of substantial modification is one of the most consequential and least-understood provisions for enterprise AI teams. Under the Act, a person who substantially modifies a GPAI model or an AI system is treated as a provider of the modified model or system, with all the obligations that provider status entails. Substantial modification is not defined with mathematical precision, and the European AI Office is expected to issue guidance on what types of changes cross the threshold. However, the Act's intent is clear: modifications that materially change the model's capabilities, risk profile, or intended use should not be used by deployers to escape provider obligations.

In practice, the activities most likely to trigger substantial modification analysis include fine-tuning a foundation model on proprietary data that materially changes its domain expertise or behavior; substantially modifying the system's intended purpose such that it is used for a high-risk application that was not within the original provider's intended use; building an application layer that is so deeply integrated with the model's outputs that it effectively constitutes a new AI system; and modifying the model's safety filters or behavioral constraints in ways that expand its capabilities beyond the provider's documented intended use.

Enterprises should document their integration architecture in enough detail to support a defensible analysis of whether their modifications constitute substantial modification under the Act. That documentation should include a description of the base model's intended use as specified by the provider, a description of all configurations, prompt templates, fine-tuning, and integration layers applied by the enterprise, an analysis of whether these modifications materially change the model's capabilities or risk profile relative to the base model, and a legal opinion on the substantial modification question where the analysis is not clear-cut. This documentation is not merely a compliance artifact: it is the foundational evidence needed to defend the enterprise's provider or deployer status in any regulatory inquiry or enforcement action.

9.6 Manager's Checklist

The following items represent the minimum set of verifiable actions that a manager responsible for AI product development or procurement should be able to confirm. Each item corresponds to a specific risk or obligation addressed in this chapter, and each should be documented in the organization's compliance records.

- Confirm whether each foundation model used in production qualifies as a GPAI model under the Act's definition and whether it has been designated as a systemic-risk model by the AI Office.
- Verify that the organization has received or obtained access to Annex XI-compliant technical documentation from each GPAI model provider and that this documentation is current with the model version in production.
- Review each foundation model vendor contract for provisions covering documentation update notifications, copyright compliance representations, training data summary access, and liability allocation for serious incidents.
- Conduct a substantial modification analysis for each deployment that involves fine-tuning, significant prompt engineering, or integration layers that materially alter the base model's behavior and obtain legal review where the analysis is not clear-cut.
- Establish a version-tracking mechanism for foundation model documentation so that compliance artifacts are updated when the model provider releases a new model version or updates the documentation.

- For deployments involving systemic-risk models, confirm that the vendor agreement includes provisions for red-team finding access and that the enterprise has an incident reporting process that meets the timeline requirements for forwarding serious incident information to the model provider.
- Complete a comparative compliance analysis before selecting between commercial and open-source foundation models, explicitly addressing the documentation, copyright, and modification obligations associated with each option.
- Assign a named owner within the compliance function for ongoing monitoring of AI Office designation changes, guidance releases, and investigation outcomes affecting the foundation models used in production.

9.7 Takeaway

The foundation model problem is fundamentally a supply chain compliance problem, and the enterprises best positioned to manage it are those that treat their foundation model vendors with the same compliance rigor they apply to regulated financial service providers or medical device component suppliers. The Act distributes obligations across the AI supply chain so that every participant, including those who neither train models nor deploy them directly to end users, must understand their precise position and maintain documentation to defend it.

For the manager who opened this chapter, uncertain about whether their product team was a provider, a deployer, or something in between: the answer depends on the facts, and the facts need to be documented now. Model version records, contract terms, prompt architecture decisions, fine-tuning logs, and modification histories are the raw material of a defensible compliance position. Organizations that establish the discipline to maintain this documentation as part of their normal engineering and procurement workflows will be able to answer regulatory questions that currently take working sessions with multiple departments to begin addressing. That operational clarity is what responsible-by-design AI development looks like in the foundation model era.

10 Designing the Compliance Function

A chief compliance officer at a global financial services firm has just received board approval to build out a dedicated AI compliance function. The firm uses AI across three major business lines: retail banking, where models are used for credit scoring and fraud detection; asset management, where models assist with portfolio analysis and trade execution; and operations, where AI tools support document processing, customer service routing, and regulatory reporting. The compliance officer has six months to present an operating model that specifies organizational design, staffing, reporting lines, tooling, and a three-year maturity roadmap. The chief risk officer favors a federated model given the firm's decentralized business structure. The chief technology officer wants a centralized technical center of excellence. The general counsel wants the function within legal. The chief compliance officer knows that all three positions reflect legitimate organizational interests and that the governance design choice will determine whether compliance is seen as a business enabler or a bottleneck for years to come.

The firm already has a data protection function, a model risk management function under the banking regulators' supervisory guidance, and an existing enterprise risk framework. The new AI compliance function cannot simply absorb these existing structures, because the EU AI Act creates obligations that extend beyond data protection and model risk. Still, it also cannot afford to duplicate work that those functions already perform. The design task is to identify gaps, build bridges, and staff the function to close both gaps without creating a bureaucratic layer that slows AI deployment across business lines.

10.1 Why It Matters

Compliance with the EU AI Act is not a documentation project. It is an organizational capability that must be built, staffed, resourced, and sustained across the entire lifecycle of every AI system the organization deploys. Documentation produced without an organizational structure to maintain and update it degrades rapidly. Risk assessments performed as one-time exercises rather than continuous monitoring activities produce a false sense of assurance. Conformity assessments completed at deployment that are never revisited as systems evolve are artifacts of historical compliance, not evidence of present compliance. The Act's requirements are ongoing, not static, and only an organizational function with clear responsibilities, sufficient authority, and adequate resources can reliably meet them.

For managers in enterprises still in the early stages of developing AI compliance programs, the organizational design decisions made now will shape the effectiveness of the compliance function for years to come. The choice between a centralized and a federated model has consequences for how quickly compliance reviews can be completed, how consistently standards are applied across business units, and how well the compliance function can recruit and retain people with the technical skills needed to understand AI systems. The reporting line decision determines whether the compliance function has the authority to slow or stop deployment when required, or whether it is positioned as an advisory body whose recommendations can be overridden by business leaders with conflicting incentives.

For managers in more mature organizations that already have AI governance structures in place, this chapter addresses the gaps that those structures typically leave: the specialized technical roles that general compliance functions do not carry, the tooling investments that make ongoing compliance operationally feasible, the metrics that signal whether the function is working, and the maturity progression from reactive, ad-hoc compliance management to a proactive, institutionalized program that meets the EU AI Act's standard for responsible-by-design AI governance.

10.2 Organizational Design Options for AI Compliance

Before addressing roles, tooling, or metrics, a compliance leader must make a foundational organizational design decision: where does the AI compliance function sit within the enterprise structure, and how does it relate to existing functions with adjacent responsibilities? The answers to these questions determine the function's authority, access to information, relationship with business units, and ability to attract and retain the people it needs.

Centralized vs. Federated Compliance Models: Tradeoffs and Fit Factors

A centralized AI compliance model concentrates all compliance expertise, authority, and accountability in a single organizational unit. This unit sets policy, performs or coordinates conformity assessments, maintains the AI inventory, reviews new deployments, and interfaces with regulators. Centralization produces consistency: all AI systems are evaluated against the same standards by people who share a common understanding of the Act's requirements, using the same tools and documentation templates. It also makes it easier to recruit specialized talent, because there is a critical mass of AI compliance work in one place rather than fractional roles distributed across business units.

A federated model distributes AI compliance responsibility across business units, with each unit maintaining embedded compliance capacity to support its own AI deployments, and a thin central function responsible for policy, standards, and cross-unit coordination. Federated models suit large, decentralized organizations in which business units operate with significant autonomy, and the pace and variety of AI deployment make centralized review a bottleneck. The risks of a federated model are inconsistency, gaps in expertise in units that lack sufficient AI compliance volume to justify a skilled specialist, and difficulty maintaining a unified AI inventory across units with different data systems and cultures.

Most organizations will adopt a hybrid model: a central function responsible for policy, standards, tooling, and high-stakes reviews, with embedded compliance partners in business units who serve as the first point of contact for new AI initiatives and maintain unit-level records. The optimal balance point between central and federated capacity depends on several fit factors: the number and diversity of AI deployments, the degree of business unit autonomy in the enterprise operating model, the maturity of existing compliance functions in individual units, the technical sophistication of business unit teams, and the organization's risk tolerance for compliance inconsistency across units. Documenting the rationale for the design choice made is itself a governance artifact: the board and regulators may ask why the organization is structured the way it is, and a reasoned design document is a stronger answer than an improvised explanation.

Diagram 8.1 - Centralized vs. Federated Compliance Models

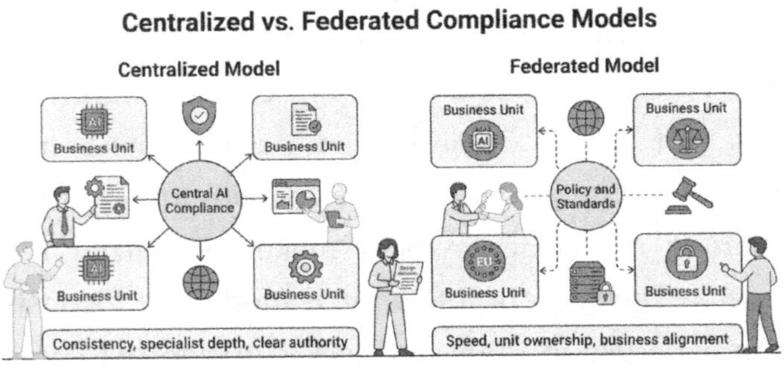

Comparing organizational models for effective AI governance.

Where AI Compliance Sits in the Enterprise Structure

The reporting line for the AI compliance function determines its character and its effectiveness in ways that are difficult to reverse once established. Each of the four main options, sitting within legal, within enterprise risk, within technology, or as a standalone function, creates different incentive structures and different relationships with the rest of the organization. Within legal, the function benefits from attorney-client privilege protections for compliance communications, direct access to regulatory counsel, and a natural home for the Act's documentation and conformity assessment requirements. The risk is that legal teams may lack the technical depth to evaluate AI system behavior, and that business units may view the function as primarily focused on liability management rather than on practical compliance enablement.

Within enterprise risk, the function benefits from existing risk frameworks, established risk appetite statements, and direct access to the board's risk committee. The risk is that AI compliance may be subordinated to financial and operational risk priorities that dominate enterprise risk agendas, and that the function may be sized according to risk function cost norms rather than the actual scale of the AI compliance task. Within technology, the function benefits from technical credibility, close relationships with engineering teams, and fast access to system documentation. The risk is that technology-based compliance functions may lack independence and face pressure to approve deployments that are not yet compliant to meet product delivery timelines.

A standalone AI compliance function, reporting directly to the CEO or to the board's audit and risk committee, offers the highest degree of independence and the clearest accountability. Still, it requires a level of organizational seniority and resource commitment that most enterprises are not yet prepared to make. In practice, the most common effective design in the short term is for AI compliance to sit within an existing compliance or risk function while maintaining a direct reporting channel to the board on matters relating to high-risk AI system governance. The key is that the reporting line must give the function sufficient authority to escalate concerns and to say no when required, regardless of where in the org chart it sits.

Diagram 8.2 - AI Compliance Reporting Line Options

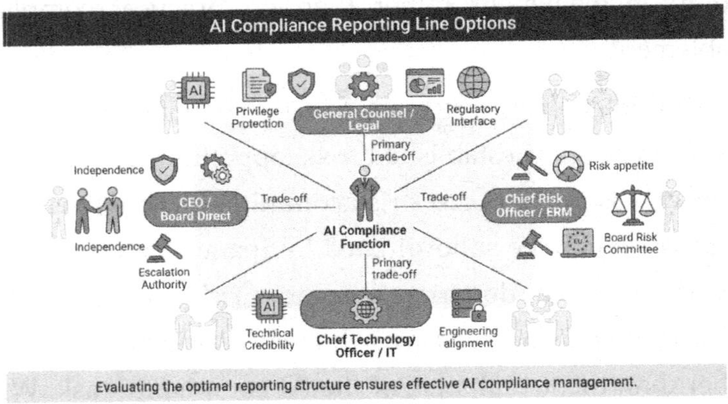

The Relationship Between AI Compliance and Adjacent Functions

Effective AI compliance does not operate in isolation. It draws on and contributes to at least three adjacent organizational functions that already exist in most large enterprises: data protection (GDPR compliance and privacy impact assessment), information security, and model risk management or model validation. Each of these functions has developed expertise, processes, and tooling that the AI compliance function can leverage, and each has interfaces with AI compliance that must be explicitly designed rather than left to develop organically.

The data protection function is the natural collaborator for AI systems that process personal data, which includes most AI systems used in HR, customer-facing applications, and healthcare. The AI Act's requirements for data governance in high-risk systems (Article 10) overlap significantly with GDPR requirements for data minimization, purpose limitation, and data quality. Rather than conducting parallel assessments, the AI compliance function and the data protection function should coordinate their review processes to produce a single integrated assessment that satisfies both frameworks. The Data Protection Impact Assessment (DPIA) process can often be extended or modified to incorporate AI Act-specific data governance requirements, avoiding duplication and reducing the compliance burden on business units.

The model risk management function, particularly in regulated financial institutions, has established processes for model validation, performance monitoring, and model inventory management that directly align with the AI Act's requirements for high-risk AI systems. The AI compliance function should map its obligations to the existing model risk management framework before designing new processes. It should document where the AI Act's requirements exceed what model risk management already covers. Building AI compliance on top of model risk management foundations is almost always more efficient than building it from scratch, and it avoids the organizational confusion of having two separate functions conducting overlapping reviews of the same AI systems.

Diagram 8.3 - AI Compliance Integration with Adjacent Functions

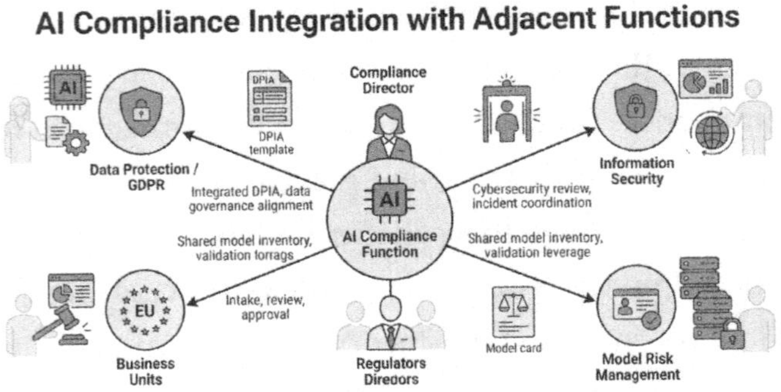

AI compliance integration with coder security and business functions.

10.3 Defining the Core Roles

The AI compliance function requires a specific set of roles that do not exist in most organizations today, and that cannot be filled simply by redeploying general compliance staff. The combination of legal, technical, and operational knowledge required to assess AI systems under the Act is unusual, and the scarcity of people who possess it creates both a recruiting challenge and a retention imperative. Understanding what roles are needed, what expertise each requires, and how they interact is the foundation of a realistic staffing plan.

The AI Compliance Lead: Responsibilities, Required Expertise, and Authority

The AI compliance lead is the function's most senior role and the individual accountable to executive leadership and the board for the organization's AI compliance posture. This person must have sufficient understanding of the EU AI Act's requirements to interpret them for the organization, sufficient authority to make consequential compliance decisions or to escalate them with credibility, and sufficient organizational credibility to be taken seriously by engineering, legal, risk, and business unit leaders who may have conflicting views about compliance priorities.

The expertise profile for an effective AI compliance lead combines three domains: regulatory knowledge covering the EU AI Act and related frameworks including GDPR, product liability, and sector-specific AI regulation; technical literacy covering AI system architecture, machine learning concepts, and data pipeline design at a level sufficient to understand system behavior and evaluate technical documentation without necessarily being an engineer; and organizational leadership covering stakeholder management, cross-functional coordination, budget management, and communication with board-level audiences. People who possess all three domains are rare, and organizations should expect to invest in developing at least one of the three in a person who arrives with the other two.

The authority dimension is non-negotiable. An AI compliance lead who lacks the authority to slow or halt a deployment that poses an unacceptable compliance risk is a compliance function in name only. The position's authority to approve, condition, or reject AI system deployments should be formally defined in a governance document, and that authority should be tested and confirmed before the function is under pressure to use it. A compliance authority that has never been exercised in a consequential case is an authority that may not hold when it is most needed.

Diagram 8.4 - AI Compliance Lead Expertise Profile

Technical Compliance Specialists: AI Engineers, Auditors, and Standards Experts

Technical compliance specialists are the people who actually examine AI systems in depth: reviewing training data practices, evaluating risk management documentation, assessing algorithmic fairness metrics, testing model behavior under adversarial conditions, and verifying that technical documentation is accurate and complete. This role requires genuine technical depth, including the ability to read and understand model documentation, evaluate claims about system performance and robustness, design testing scenarios, and interpret monitoring data from deployed systems.

In most organizations, technical compliance specialists will be recruited from two adjacent talent pools: AI engineers who have developed an interest in governance and regulatory affairs, and compliance auditors who have developed technical skills through experience in regulated industries such as financial services or medical devices. Both pathways produce candidates with significant strengths and significant gaps: the AI engineer may lack regulatory knowledge and audit methodology, while the compliance auditor may lack the technical depth to evaluate AI system architecture independently. Building a team that includes both types and creating structured learning programs to close the gaps in each is the most practical approach in the current talent market.

Standards' expertise is a specific technical skill the function needs. It is difficult to develop quickly: the ability to read and apply harmonized standards, understand certification procedures, interface with notified bodies, and translate standards requirements into internal engineering processes. As the European AI Office develops harmonized standards for the Act and international standards bodies produce AI governance frameworks, the organization needs at least one person to track these standards, interpret their implications, and advise engineering and product teams on achieving conformity. This is a niche role, but it is a critical one for any organization that will undergo third-party conformity assessment.

Diagram 8.5 - Technical Specialist Role Pathways

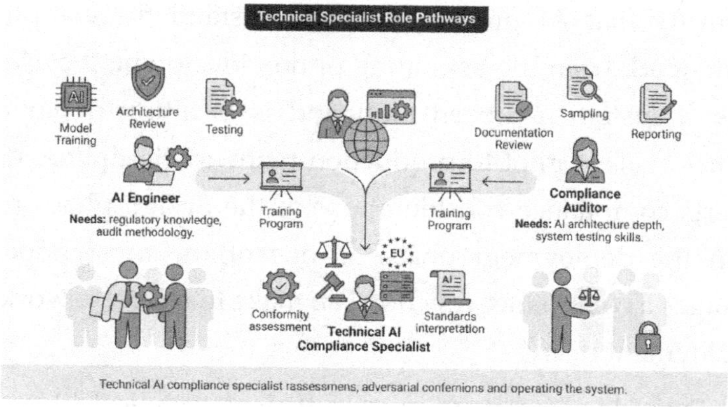

Business-Facing Compliance Partners Embedded in Product and Operations Teams

Business-facing compliance partners, sometimes called AI governance liaisons or embedded compliance advisors, are the function's interface with the teams that actually build and deploy AI systems. These individuals sit within or adjacent to product, operations, or engineering teams and serve as the first point of contact when a new AI use case is proposed, when a deployed system raises a performance concern, or when a business unit needs guidance on the compliance implications of a design decision. Their effectiveness depends on being seen as knowledgeable, accessible, and mission-aligned rather than as gatekeepers or part of an approval bureaucracy.

The embedded compliance partner role is different from the technical specialist role in emphasis. Technical specialists go deep on system evaluation; embedded partners go broad on early engagement. The partner's value is preventing compliance problems from developing in the first place by ensuring that AI project teams understand the compliance framework from the beginning of the development cycle, not after a system has been built and is awaiting deployment approval. This pilot-to-production pathway discipline, which inserts compliance considerations at the design phase rather than the deployment phase, is one of the most impactful changes a compliance function can make to reduce rework and deployment delays.

Embedded partners need sufficient technical literacy to participate credibly in product design conversations, sufficient regulatory knowledge to identify the compliance implications of design choices, and sufficient interpersonal skills to maintain productive relationships with teams that may sometimes view compliance requirements as obstacles. The selection of embedded partners is as important as the selection of technical specialists: the wrong person in this role can damage the compliance function's relationship with the business, while the right person becomes a trusted advisor whose guidance is sought proactively rather than reluctantly.

Diagram 8.6 - Embedded Compliance Partner Engagement Lifecycle

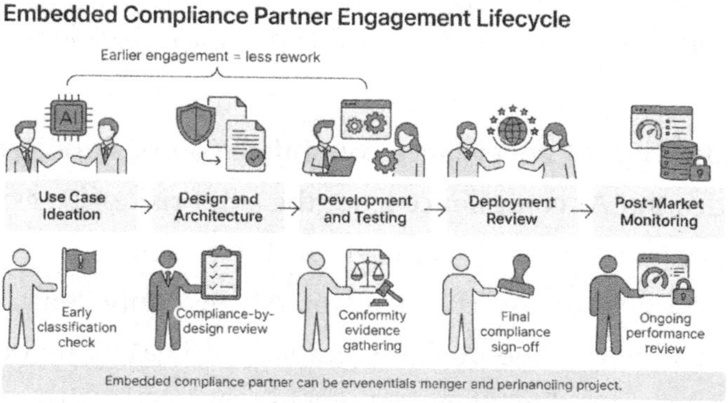

10.4 Cross-Functional Coordination Mechanisms

AI compliance cannot be managed within a single organizational silo. The Act creates obligations that require input from legal, technology, procurement, human resources, and business operations, and it creates decision points that affect all of these functions. Effective coordination mechanisms, including governance committees, escalation pathways, and cross-functional working groups, are the connective tissue that make a distributed compliance effort function as a coherent program rather than a set of parallel, disconnected activities.

AI Governance Committees: Composition, Cadence, and Decision Rights

An AI governance committee is the senior-level body responsible for making consequential decisions about the organization's AI compliance posture: approving the AI inventory classification framework, setting risk appetite for AI deployments, reviewing high-stakes conformity assessments, deciding on resource allocation for the compliance function, and escalating significant compliance concerns to the board. The committee's authority and composition must be defined precisely, because a governance committee without clear decision rights is a discussion forum, not a decision-making body.

Effective AI governance committee composition typically includes the AI compliance lead, the chief risk officer or their delegate, the general counsel or their delegate, a senior technology leader (CTO, head of AI, or equivalent), and a representative of the business units most heavily involved in AI deployment. For organizations in regulated industries, the chief compliance officer and a representative of the internal audit function should also participate. The committee should meet on a regular cadence, typically monthly or quarterly, depending on the pace of AI deployment, with provisions for ad hoc meetings when significant compliance events or decisions arise.

The decision rights framework for the committee should distinguish between decisions that require committee approval (new high-risk AI system deployments, changes to the risk classification framework, engagement with regulators on compliance matters, material changes to the AI compliance budget), decisions that the AI compliance lead can make independently and report to the committee, and decisions that can be made at the business unit level with committee policy guidance. Making this framework explicit in a governance charter and reviewing it at least annually prevents the ambiguity that can lead to either committee overload or decisions made without appropriate oversight.

Diagram 8.7 - AI Governance Committee Structure

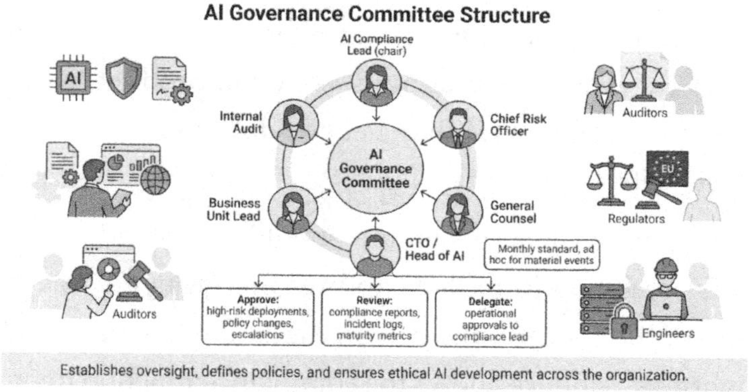

Escalation Pathways from Product Teams to Compliance Leadership

A compliance function that cannot receive clear, timely escalations from the teams it serves will always be reactive, learning about compliance problems after they have become crises rather than while they can still be addressed. Escalation pathways must be formally defined, communicated to all AI project teams, and tested periodically to ensure they function as intended. The pathway should specify which situations trigger escalation, who receives the escalation, the expected response timeline, and what documentation accompanies it.

Situations that should trigger mandatory escalation include: a product team's discovery that a deployed AI system is producing outputs that differ materially from expected behavior; a decision by an external vendor to change a model or service in ways that affect a deployment's compliance posture; a user complaint or incident report suggesting harm attributable to AI system behavior; a regulatory inquiry or investigation related to an AI system; and a proposed design change that the embedded compliance partner believes may cross a risk threshold that the governance committee should review. Making escalation straightforward and consequence-free for the person escalating is essential: product teams will not escalate concerns if they believe doing so will slow their projects without benefit.

Escalation documentation should be standardized. A simple escalation form that captures the system involved, the nature of the concern, the person escalating, the date, and any supporting evidence allows the compliance function to process escalations consistently and to maintain a log that demonstrates to regulators that the function is operating as designed. The log itself is a compliance artifact: in an enforcement context, the ability to demonstrate that escalations were received, processed, and resolved is evidence that the compliance function is operational rather than merely organizational.

Diagram 8.8 - Compliance Escalation Pathway

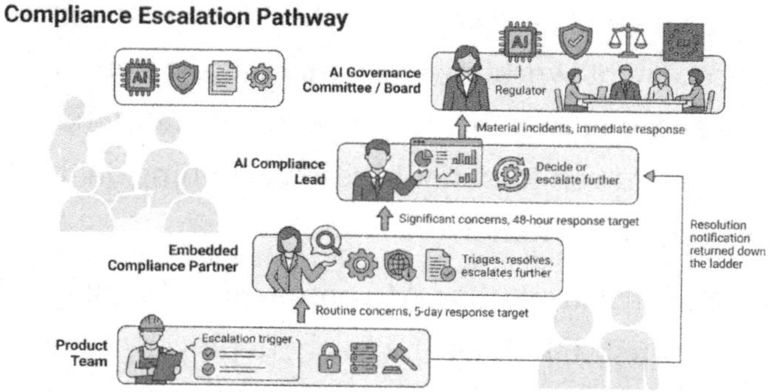

Streamlined process for efficient and compliant AI issue resolution.

Coordinating with Legal, Procurement, and Human Resources on AI Compliance Obligations

Three functions outside the AI compliance organizational unit have specific and non-delegable responsibilities in the AI compliance program: legal, procurement, and human resources. Legal is responsible for contract review and negotiation, regulatory correspondence, enforcement response preparation, and legal opinions on ambiguous compliance questions. The AI compliance function's relationship with legal must be structured so that legal capacity is available when needed without creating bottlenecks, and so that legal advice on compliance matters is documented in a form that supports audit inquiries.

Procurement is responsible for the vendor due diligence and contract negotiations that determine the quality of compliance information flowing from foundation model providers and other AI vendors. Procurement teams that are not trained to identify AI Act-relevant contract provisions will negotiate contracts that leave compliance gaps, even when the compliance function has published standards for what vendor contracts must contain. The AI compliance function should develop a vendor contract review checklist and integrate it into the procurement workflow as a mandatory step for any AI-related vendor engagement. Procurement staff who regularly engage with AI vendors should receive training on the AI Act's vendor obligation framework.

Human resources has AI compliance obligations in two directions. First, HR is itself a likely deployer of high-risk AI systems, using AI tools for recruitment, performance management, and workforce analytics in ways that the Act classifies as high-risk. The AI compliance function must ensure that HR's AI deployments meet the same standards as those of any other business unit. Second, HR manages the hiring processes for AI compliance roles, and the compliance function's ability to recruit effectively depends on HR understanding the unusual expertise requirements of these roles and the compensation levels needed to attract qualified candidates in a competitive market for AI governance talent.

10.5 Budgeting, Tooling, and Maturity for Compliance

An under-resourced compliance function will operate in permanent triage mode, prioritizing only the most visible risks while structural gaps accumulate. A function that is resourced without discipline will accumulate activities and reports without improving the organization's actual compliance posture. The discipline of compliance investment planning, tooling selection, and maturity tracking ensures that resources produce measurable outcomes rather than organizational activity.

Sizing the Compliance Investment: A Framework for Estimation

There is no industry consensus on what AI compliance should cost as a percentage of AI operating budget, and the early state of the regulatory environment means that benchmarking data is sparse. However, a practical framework for estimating the required investment begins with the organization's AI inventory: the number of AI systems in production, their risk classification, and the intensity of compliance obligations for each classification. A small number of high-risk systems may require more compliance investment than a large number of low-risk systems, because they require conformity assessments, technical documentation, ongoing monitoring, and possibly a third-party audit.

A rough sizing methodology identifies three cost drivers: headcount for the compliance function itself, tooling and infrastructure costs for the compliance technology stack, and engagement costs for external advisors, including legal counsel, standards experts, and notified bodies for third-party conformity assessments. For an organization with five to ten high-risk AI systems and a broader portfolio of limited and minimal-risk systems, a compliance function of four to six dedicated staff (compliance lead, two technical specialists, one or two embedded partners, and an administrative coordinator) plus a compliance tooling budget and a reserve for external advisory costs represents a defensible baseline investment.

The business case for AI compliance investment should be framed in terms of risk avoided, not just compliance achieved. The EU AI Act's penalties for non-compliance with high-risk system obligations range from 30 million euros to 6% of global annual turnover for the most serious violations. A compliance function that prevents a single material enforcement action more than justifies its annual operating cost. The business case should also reference the commercial benefits of compliance: access to regulated markets, customer trust, reduced insurance costs, and the ability to participate in public procurement processes that require AI compliance certification.

Diagram 8.9 - Compliance Investment Sizing Framework

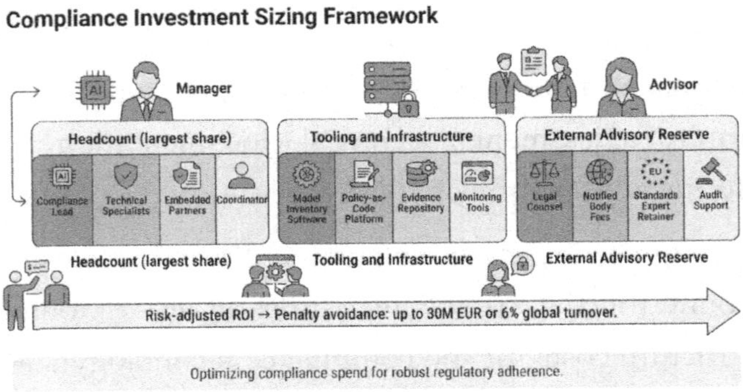

The Compliance Tooling Stack: Inventory, Policy, and Evidence

A compliance function that relies primarily on spreadsheets and shared drives to manage its AI inventory, documentation, and review workflows will find that the system's limitations become binding constraints as the AI portfolio grows. The compliance tooling stack, the set of software tools used to manage AI compliance operations, must be selected with both current needs and future scale in mind. The core components of a compliance tooling stack include a model inventory system, a policy management platform, an approval workflow tool, and an evidence repository.

The model inventory system is the most foundational component. It is a structured record of every AI system in the organization's portfolio, including the system's risk classification, the business unit responsible for it, the version currently in production, the date of the last conformity assessment, the system's current monitoring status, and any open compliance findings. The inventory must be kept current by integrating with the organization's software development lifecycle processes: systems added to production should automatically appear in the inventory, and version changes should trigger notifications that prompt compliance reviews. An inventory maintained manually, separate from engineering systems, will quickly drift out of sync with reality.

Policy-as-code tools allow compliance policies to be expressed in machine-readable formats that can be evaluated automatically against system configurations, infrastructure settings, and deployment parameters. Rather than relying on human review to verify that every AI system deployment meets policy requirements, a policy-as-code approach enables automated checks that flag non-compliant configurations before deployment. Evidence repositories provide a structured, audit-ready record of the documentation, test results, review decisions, and approval logs that demonstrate compliance for each system. Organizing evidence in a way that maps to the Act's specific requirements, rather than in a format that made sense at the time of collection, makes the audit preparation process dramatically faster and reduces the risk of gaps being discovered only when an auditor asks for specific documentation.

Diagram 8.10 - AI Compliance Tooling Stack Architecture

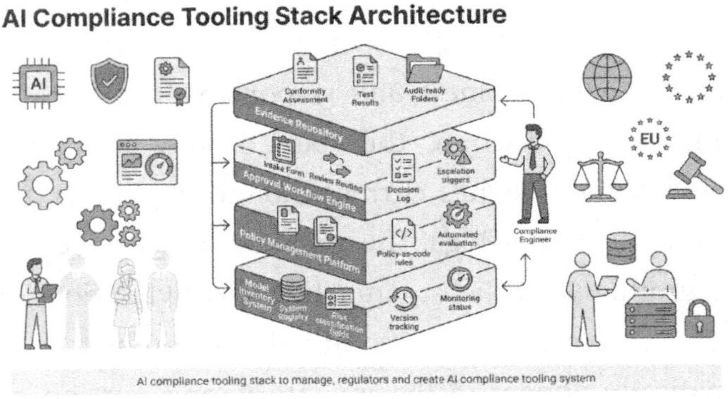

Maturity Progression from Ad-Hoc to Institutionalized Compliance

Most organizations that are beginning to address the EU AI Act compliance are operating at an ad-hoc level: compliance activities are performed reactively, driven by specific deadlines or pressure events, without a consistent methodology or a maintained record of prior work. Moving from ad-hoc to institutionalized compliance requires progressing through several recognizable maturity stages, each of which builds the foundation for the next.

The first stage above ad-hoc is documented: the organization has written policies for AI compliance, a defined scope for the compliance program, and a process for classifying AI systems. At this stage, compliance activities are consistent in approach, even if they are not yet automated or comprehensive. The second stage is managed: the organization has a complete AI inventory, conducts conformity assessments for all high-risk systems, maintains ongoing monitoring of deployed systems, and has a governance committee that regularly reviews compliance status. At this stage, compliance is a sustained operational activity rather than a periodic project. The third stage is optimized: the organization uses data from its compliance activities to continuously improve its processes, has compliance tooling that reduces manual effort and improves consistency, tracks metrics that indicate the health of the compliance function, and maintains a compliance posture that regulators would regard as exemplary.

The metrics of a healthy compliance function at the managed and optimized stages include: the percentage of AI systems in the inventory with a current (less than twelve months old) risk classification; the average time from AI system deployment to conformity assessment completion for high-risk systems; the percentage of high-risk systems with current monitoring plans and non-expired monitoring results; the number of compliance findings identified at pre-deployment review versus post-deployment (the ideal ratio is high pre-deployment, indicating that compliance review is effective at catching issues before they reach production); and the time from regulatory guidance publication to internal policy update reflecting that guidance. These metrics, tracked over time and reported to the governance committee, provide an operational view of whether the compliance function is working, not just whether the right activities are being performed.

10.6 Manager's Checklist

The following items represent the organizational decisions and operational steps that a manager building or maturing an AI compliance function should be able to confirm. Each item is verifiable against a specific document, decision record, or operational artifact.

- Document the rationale for the organizational design decision (centralized, federated, or hybrid) and the reporting line chosen, including the authority the AI compliance lead holds to approve, condition, or reject AI system deployments.
- Define the decision rights of the AI governance committee in a formal governance charter, specifying which decisions require committee approval and at what cadence the committee meets.
- Establish a cross-functional coordination protocol with legal, procurement, and HR that specifies each function's AI compliance obligations and the interfaces between them and the AI compliance function.
- Publish and distribute a mandatory escalation pathway document to all AI project teams, defining the situations that require escalation and the response timelines for each escalation tier.
- Complete a tooling needs assessment and select at minimum model inventory system and an evidence repository, with a plan to ingest all current AI systems within ninety days of deployment.
- Develop a compliance investment business case that quantifies the risk-adjusted return on the compliance function investment, referencing the Act's maximum

penalty provisions and the commercial benefits of documented compliance.

- Define the maturity metrics that the compliance function will track and report to the governance committee, establishing baseline values and twelve-month improvement targets for each.
- Publish a procurement checklist for AI vendor contracts that embeds Act-compliant documentation and liability allocation requirements into the standard vendor onboarding workflow.

10.7 Takeaway

Designing the AI compliance function is one of the most consequential organizational decisions enterprise leaders will make as the EU AI Act comes into full effect. The decisions made now about reporting lines, role definitions, committee authority, and tooling investments will shape the function's effectiveness for years, and they are difficult to reverse once embedded in organizational structure and culture. The manager who approaches this design with the same rigor applied to any significant organizational build will produce a function capable of meeting the Act's demands; the manager who defers the design to improvise later will find that improvisation costs more in regulatory risk and rework than the upfront investment would have required.

The most important design principle is that the compliance function must have genuine authority matched to real accountability. A compliance function that is well-staffed, well-tooled, and well-connected to the business but that lacks the authority to stop a non-compliant deployment is not a compliance function. It is an advisory service that regulators will not recognize as satisfying the Act's governance requirements. Build the function with authority from the beginning, test that authority before it is needed in a high-stakes situation, and maintain its independence from the commercial pressures that will inevitably be brought to bear on every significant compliance decision.

11 Governing Through Change

A board chair at a multinational industrial technology company has just received a quarterly compliance report showing that the company's EU AI Act compliance program is on track against all documented milestones. The program team has completed risk classifications for the company's twenty-three AI systems, produced conformity assessments for the seven classified as high-risk, and established post-market monitoring protocols for all deployed systems. The compliance function is staffed, the governance committee is meeting on schedule, and the documentation is audit-ready. The board chair asks one question: 'What happens when the requirements change?' The compliance lead pauses. The milestones were built against the Act as published. The delegated acts that will add specificity to several key provisions have not yet been adopted. The harmonized standards that will define technical conformity are still in development. The AI Office has published draft guidance that may or may not reflect the final position. And three of the company's major markets, the United States, the United Kingdom, and Canada, are developing their own AI regulatory frameworks that may impose requirements that diverge from the EU Act in material ways.

The board chair's question reveals a structural gap in the compliance program. The program was designed to achieve compliance at a point in time. It was not designed to maintain compliance across time. The investments made in building the program created real value. Still, they do not, by themselves, constitute a governance capability for a regulatory environment that will continue to evolve for the foreseeable future. What the company now needs is not a project plan but a governance operating model that treats regulatory change as a continuous operational condition rather than an exceptional event.

11.1 Why It Matters

The EU AI Act is not a finished regulatory product. It is a framework statute whose practical content will be filled in over the years through delegated acts, implementing acts, codes of practice, harmonized standards, and AI Office guidance. Organizations that achieve compliance against the Act as written today will find that the compliance target moves repeatedly, and that maintaining compliance requires an ongoing organizational investment that looks more like a permanent operational function than a closed implementation project. The companies that build governance structures capable of absorbing regulatory change will compound their compliance investment over time; the companies that treat compliance as a project will face repeated cycles of assessment, remediation, and re-assessment every time the regulatory environment shifts.

The global dimension of AI regulation adds further complexity. Most large enterprises operate across multiple jurisdictions that are simultaneously developing AI governance frameworks. The EU AI Act is the most comprehensive and most immediately enforceable of these frameworks, but it is not the only one that matters. US state laws, including Colorado's AI Act, Texas and Illinois employment AI requirements, and a growing number of state-level proposed bills, impose obligations that may conflict with

EU requirements in specific contexts. The UK's sector-specific AI governance approach, Canada's Artificial Intelligence and Data Act, Brazil's AI framework bill, Singapore's Model AI Governance Framework, and China's series of AI-specific regulations each add compliance obligations that must be managed alongside the EU Act. An enterprise that manages each jurisdiction's requirements in isolation will face compounding complexity; an enterprise that builds a coherent global compliance architecture that prioritizes the most demanding applicable requirements will find that the additional cost of global compliance over EU-only compliance is manageable.

The strategic value of building an adaptive governance capability extends beyond compliance itself. Organizations that develop strong regulatory intelligence functions, that build relationships with standards bodies and regulators, and that participate in shaping the guidance and standards that will govern their industry are not merely surviving the regulatory environment. They are influencing it. The manager who reads this chapter as a checklist of compliance obligations is capturing part of its value; the manager who reads it as a guide to building a regulatory engagement capability that creates competitive advantage is capturing the rest.

11.2 The Regulatory Trajectory of AI Governance

Before an organization can build governance structures that adapt to regulatory change, it must understand the architecture of the regulatory change process itself. The EU AI Act's development pathways, the mechanisms through which its requirements will evolve, and the institutions that will drive that evolution are not incidental details. They are the operational context that the compliance function must monitor, engage with, and respond to continuously.

How the EU AI Act Is Expected to Evolve Through Delegated Acts and Implementing Measures

The EU AI Act grants the European Commission the authority to adopt delegated acts and implementing acts that add technical specificity to the framework statute's requirements. Delegated acts are legislative measures that supplement or amend non-essential elements of the Act; implementing acts are measures that establish the conditions for the uniform application of the Act across member states. Both types of acts are adopted through processes that involve the Commission, the Council of the EU, and the European Parliament, and both are subject to public consultation periods during which organizations can submit comments and positions.

The areas most likely to be addressed through delegated acts and implementing measures include: updates to the list of high-risk AI system categories in Annex III as new application domains develop; updates to the technical requirements for specific high-risk categories; the procedures for notified body designation and oversight; the format and content requirements for EU declarations of conformity; and technical standards for post-market monitoring and serious incident reporting. Each of these updates represents a potential change to an organization's compliance obligations, and the organization's compliance function must have a process for identifying when a relevant act is under development, tracking its progress, assessing its implications for the organization's AI systems, and updating internal policies and documentation when the act enters into force.

The timeline for delegated and implementing acts under the EU AI Act is uncertain, and the compliance function should not plan against specific dates. Instead, the function should establish a regular scanning process that reviews the Official Journal of the European Union, the AI Office's published work program, and the Commission's Better Regulation portal for new and developing legislative instruments. When a relevant instrument is identified, the function should assess whether it creates a compliance gap, escalate the assessment to the governance committee, and initiate a policy update process proportionate to the significance of the change.

Diagram 9.1 - EU AI Act Legislative Evolution Pathways

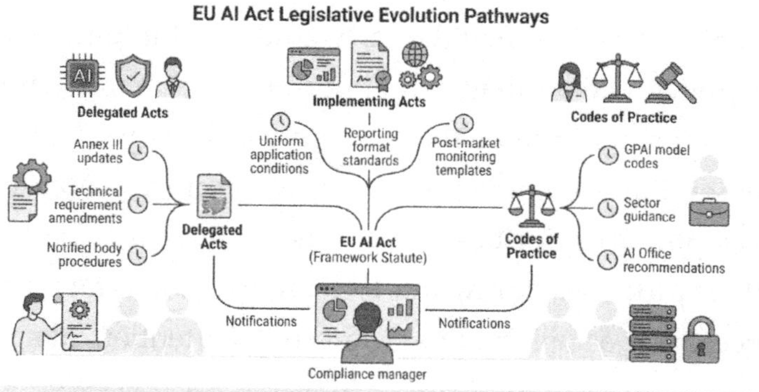

Continual evolution of the AI Act framework through distinct regulatory pathways.

The European AI Office: Role, Authority, and Enforcement Posture

The European AI Office was established within the European Commission to serve as the central supervisory body for GPAI models and to coordinate AI governance across member states. Its authority includes direct supervision of GPAI model providers and systemic-risk models, the development of guidance and codes of practice, coordination with national competent authorities responsible for supervising high-risk AI system deployments, and, in cases of serious risk, the ability to take urgent intervention measures, including requiring the market withdrawal of a non-compliant model.

The AI Office's enforcement posture during its early operational period is expected to focus on building the regulatory infrastructure (codes of practice, evaluation methodologies, and harmonized standards references) rather than on punitive enforcement. However, this does not mean that enforcement is not coming. The Act creates substantial penalties for non-compliance, and the Commission has both the authority and the political mandate to enforce them. Organizations that treat the early guidance-building period as a reprieve rather than an opportunity to build compliant systems are misjudging the regulatory trajectory. The AI Office is observing the market, building enforcement capacity, and developing its priorities in real time. Engaging constructively with its processes during this formative period is far preferable to becoming a subject of its enforcement attention later.

For enterprise compliance functions, the AI Office is both a regulatory authority and an information resource. The Office publishes guidance, codes of practice, and technical documentation that are among the most authoritative interpretive resources available for the Act's requirements. The compliance function should establish a formal process to review AI Office publications upon release, assess their implications for the organization's compliance posture, and incorporate relevant guidance into internal policies and procedures. The time between AI Office publication and internal policy update should be tracked as a compliance metric: a long lag indicates that the horizon-scanning function is not operating effectively.

Diagram 9.2 - European AI Office Authority Structure

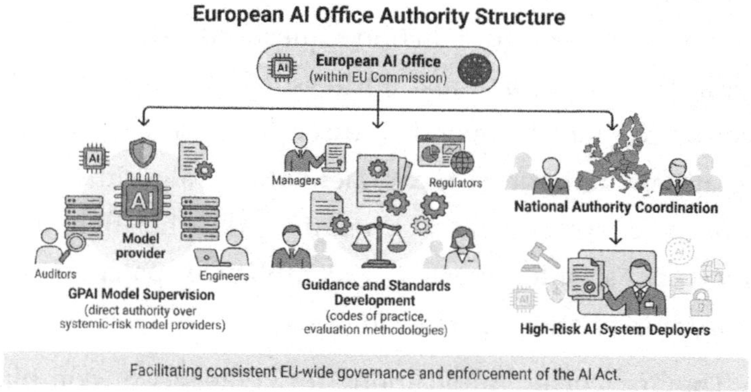

Anticipating Where Global AI Regulation Converges and Diverges

The global AI regulatory landscape is developing rapidly but unevenly. Several broad areas of convergence are emerging: most major regulatory frameworks share a risk-based approach that imposes more demanding obligations on AI used in high-stakes decisions about individuals; most frameworks include some form of transparency or explainability requirement; and most frameworks include human oversight requirements for high-risk deployments. These convergent elements enable a compliance architecture that satisfies multiple frameworks by addressing their shared core requirements.

The areas of divergence are also significant. The EU AI Act's ex-ante conformity assessment requirement, which mandates that high-risk AI systems demonstrate compliance before market placement, is more demanding than the primarily ex-post enforcement approach used in most US state laws and in the UK's sector-specific guidance. The EU Act's extraterritorial scope, which applies to providers and deployers anywhere in the world whose systems affect EU residents, creates jurisdiction questions that US-centric compliance programs must address. China's approach to AI governance, which operates through a series of specific regulations targeting recommendation algorithms, deep synthesis (deepfakes), and generative AI rather than a single framework statute, creates a compliance patchwork that requires jurisdiction-specific analysis.

The practical implication for enterprise compliance programs is that global coverage requires a layered architecture: a base layer that satisfies the most demanding applicable requirements (generally the EU AI Act for organizations with EU market exposure), supplemented by jurisdiction-specific modules that address the requirements of individual markets where those requirements differ from or exceed the EU baseline. Building this architecture requires a systematic mapping of applicable laws in each jurisdiction where the organization operates, identifying requirements that exceed the EU baseline, and integrating those requirements into the relevant system-level compliance documentation. This mapping should be reviewed at least annually and whenever a significant change in an applicable jurisdiction's regulatory framework is identified.

Diagram 9.3 - Global AI Regulation Convergence Map

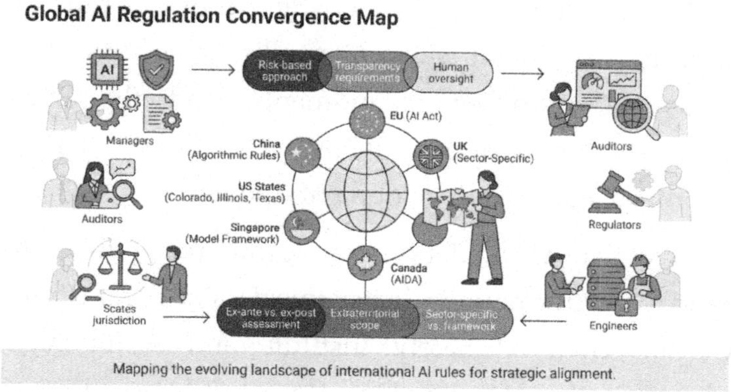

Mapping the evolving landscape of international AI rules for strategic alignment.

11.3 Horizon Scanning and Regulatory Intelligence

Horizon scanning is the organizational capability to identify and assess regulatory developments before they become binding obligations, giving the compliance function time to adapt rather than react. Building this capability requires systematic monitoring of multiple information sources, processes for translating regulatory signals into internal compliance planning, and the organizational discipline to act on early intelligence rather than waiting for regulatory certainty that may not arrive until an obligation is imminent.

Building a Regulatory Monitoring Capability: Sources, Processes, and People

A regulatory monitoring capability requires three components: a curated set of information sources, a process for regularly reviewing and synthesizing information from those sources, and a person or team responsible for maintaining the capability. The information sources for EU AI Act compliance monitoring include the Official Journal of the European Union for adopted delegated and implementing acts, the AI Office's website for guidance publications and codes of practice, the European Parliament and Council's legislative tracking systems for proposed amendments, and the European Standards Organizations (CEN, CENELEC, ETSI) for harmonized standard development updates.

For global AI regulation monitoring, the source set must expand to include national legislative databases for US state AI laws, the UK's AI Safety Institute and sector regulator publications, the Canadian Department of Innovation Science and Economic Development for AIDA developments, Singapore's IMDA and PDPC for model governance framework updates, and relevant international bodies including the OECD, ISO, and the Global Partnership on AI for international convergence initiatives. Managing this information flow requires either a dedicated monitoring tool or a structured division of responsibility among compliance team members for tracking specific jurisdictions.

The monitoring process should produce a regular regulatory intelligence report, ideally monthly, that summarizes new and developing regulatory instruments, assesses their relevance to the organization's AI systems and compliance posture, and recommends internal actions where relevant. This report should be reviewed by the AI compliance lead and shared with the AI governance committee at each meeting. The report serves both an operational function (keeping the compliance function informed) and a governance function (demonstrating to the board and to regulators that the organization is actively monitoring its regulatory environment rather than passively waiting for obligations to become visible).

Diagram 9.4 - Regulatory Intelligence Monitoring System

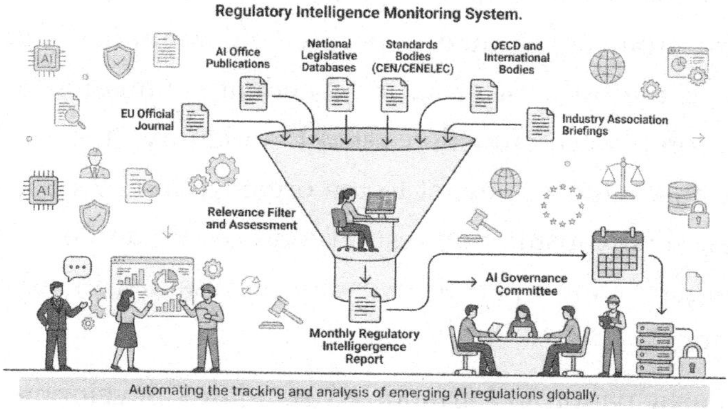

Engaging with the Regulatory Process: Consultations, Industry Associations, and Standards Bodies

Horizon scanning is a passive capability: it tracks what regulators are doing. Active engagement with the regulatory process is a strategic capability: it enables organizations to influence what regulators do, at least at the margins, and to develop regulatory relationships that provide advance notice of developing requirements and informal guidance on compliance questions. Both capabilities are valuable, but active engagement requires resources and organizational will that passive scanning does not.

The most accessible form of regulatory engagement is participation in public consultations. The European Commission and the AI Office publish consultation documents on proposed regulatory measures and invite stakeholder responses. Submissions to public consultations are public records that demonstrate the organization's engagement and allow it to advocate for interpretations of the Act's requirements that are technically sound and operationally feasible. Many organizations submit through industry associations rather than independently, which allows them to benefit from aggregated expertise and amplified influence without the full cost of an independent submission process. The compliance function should identify the industry associations most relevant to the organization's AI use cases, review their consultation response processes, and ensure that the organization's perspectives are represented in submitted positions.

Participation in harmonized standards development is a higher-resource but higher-influence form of regulatory engagement. The European Standards Organizations develop harmonized standards that, once referenced in the Official Journal, provide a presumption of conformity with the Act's requirements. Organizations that participate in these standards development processes can shape the technical requirements that will define conformity for their sector, develop advanced knowledge of what conformity will require before standards are finalized, and build relationships with the technical experts and regulators who are shaping the standards. For organizations with significant AI compliance exposure, membership in relevant technical committees of CEN, CENELEC, or ISO is a strategic investment, not a marginal one.

Diagram 9.5 - Regulatory Engagement Ladder

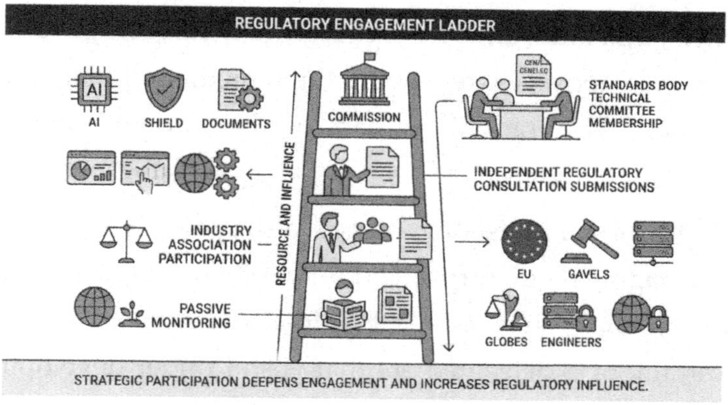

Translating Regulatory Signals into Internal Compliance Planning

Regulatory intelligence has no value unless it produces internal action. The process of translating regulatory signals into compliance planning decisions is the most operationally important step in the horizon-scanning capability, and it is the step most commonly skipped or executed informally. A formal translation process ensures that regulatory developments are assessed consistently, that their implications are escalated appropriately, and that the compliance function's response is documented to demonstrate due diligence.

The translation process should begin with a relevance assessment: does regulatory development apply to any of the organization's AI systems, operations, or markets? If the development is relevant, the next step is an impact assessment: what obligations does it create or modify, and which AI systems or processes are affected? The impact assessment should produce a gap analysis comparing the current compliance posture against the new or modified requirements and identifying the specific actions needed to close any gaps. The gap analysis feeds into a remediation plan with assigned owners, timelines, and resource requirements. The governance committee approved reviewing the remediation plan, and its execution was tracked as part of the normal compliance management cadence.

The formal translation process also provides a mechanism for managing regulatory uncertainty. When a regulatory signal is ambiguous, such as draft guidance whose final form is uncertain, or a delegated act that is still under development, the translation process should produce a contingency assessment: what is the range of plausible outcomes, and what compliance posture satisfies the most demanding plausible outcome? Taking a position that complies with the most demanding plausible interpretation of an uncertain regulatory development is generally more cost-effective than waiting for certainty and then remediating against a determined requirement under time pressure.

Diagram 9.6 - Regulatory Signal to Compliance Action Pipeline

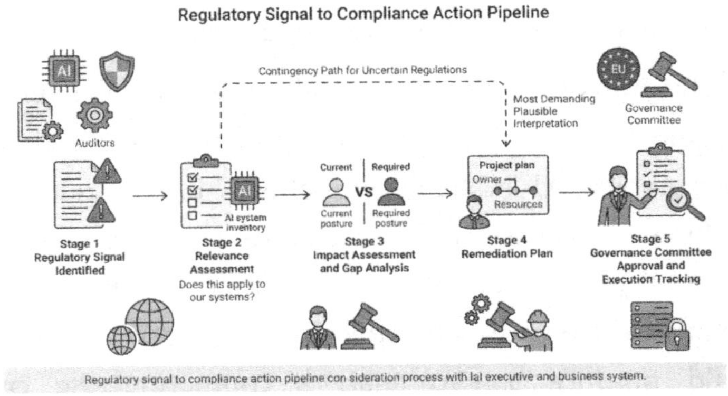

11.4 Adaptive Governance: Compliance That Moves with the Law

Horizon scanning and regulatory engagement produce intelligence. Adaptive governance is the organizational capability to act on that intelligence by modifying compliance processes, documentation, and risk postures in response to regulatory change without requiring a wholesale redesign of the compliance program each time. Building adaptability into the compliance function from the outset is far less costly than retrofitting it into a function designed for a static regulatory environment.

Designing Compliance Processes That Can Absorb Regulatory Change

The most adaptive compliance processes share a common design principle: they are structured around the underlying objectives of the regulatory requirement rather than the specific form of the requirement as written at a particular point in time. A conformity assessment process designed to verify that a specific technical standard is met is brittle: when the standard changes, the process must be redesigned. A conformity assessment process designed to verify that the AI system's risk management, data governance, human oversight, and transparency mechanisms meet the objectives of the Act is more durable: when the specific standards or technical requirements change, the assessment can be updated by modifying the technical criteria against which the objectives are evaluated, without rebuilding the assessment process itself.

Process adaptability also requires modular documentation structures. Compliance documentation organized as a single integrated document is difficult to update when any element of the regulatory framework changes, because every change requires a complete document revision. Documentation organized as a set of discrete, individually versioned artifacts, such as a risk management plan, a data governance statement, a human oversight protocol, and a transparency policy, each maintained and updated independently, can be revised element by element as regulatory requirements evolve. The modular structure also makes it easier to map documentation to specific regulatory requirements, which simplifies the audit preparation process.

Embedding change capacity in compliance processes requires that process owners receive regular training on the regulatory developments that affect their processes, that process documentation is reviewed at a defined cadence (at minimum annually, and more frequently when major regulatory changes are anticipated), and that the compliance tooling stack is capable of tracking process versions and providing an audit trail of when processes were updated and what prompted each update. A compliance process without a version history cannot be defended in an enforcement context: the ability to show that a process was current at the time of a specific compliance decision requires knowing precisely which version of the process was in effect at that time.

Diagram 9.7 - Modular Compliance Documentation Architecture

MODULAR COMPLIANCE DOCUMENTATION ARCHITECTURE

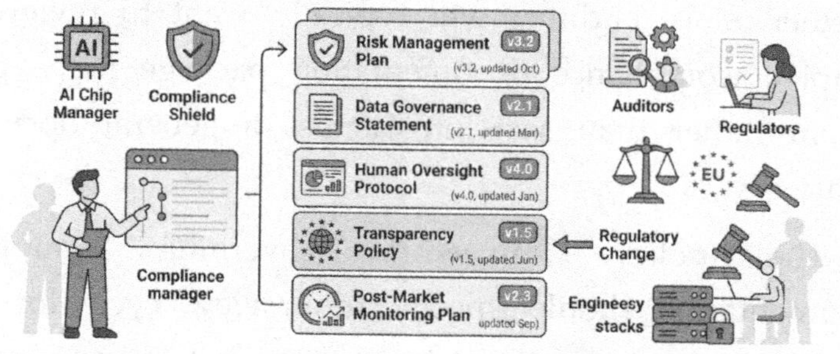

Discrete updates isolate regulatory impacts and streamline maintenance.

Version Control for Compliance Documentation and Policy Frameworks

Version control for compliance documentation is not a technical nicety: it is a regulatory necessity. The EU AI Act requires organizations to maintain up-to-date technical documentation for deployed AI systems. When regulators or auditors examine that documentation, they will want to know not only what the current version says, but also what version was in effect at specific historical dates, what changed between versions, and who authorized each change. A compliance documentation system that cannot produce this information is, from a regulatory perspective, an uncontrolled documentation system.

Effective version control for compliance documentation requires a defined versioning scheme (typically major versions for substantive changes and minor versions for editorial corrections), a change log for each document that records the nature of the change, the regulatory or operational trigger for the change, the date of the change, and the approver, and a document repository system that stores all historical versions in a retrievable format. The repository should be organized so that documents related to a specific AI system can be retrieved together, since auditors will typically want to review the complete compliance documentation package for a given system rather than searching across a general document library.

The policy framework governing compliance documentation should specify the review cycle for each document category, the triggers for out-of-cycle review (regulatory change, system modification, incident, or escalation from an embedded compliance partner), the approval authority required for each type of update, and the retention period for superseded versions. These specifications should themselves be documented in a compliance documentation management policy, which is subject to the same version control requirements as the documents it governs. This self-referential requirement is not bureaucratic recursion: it is evidence that the organization's commitment to documented compliance is itself documented.

Diagram 9.8 - Compliance Document Version Control System

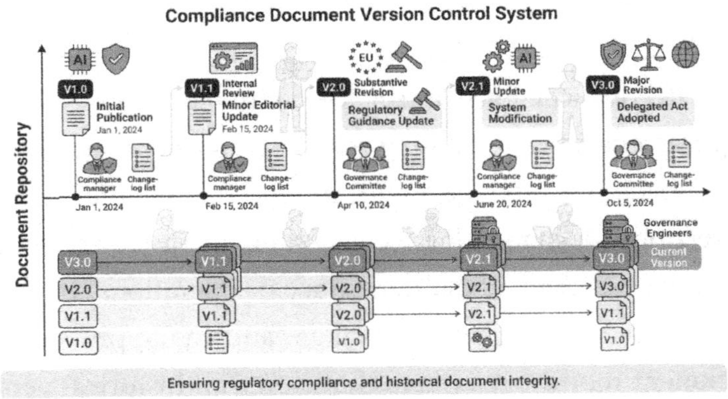

Governance Review Cycles: When and How to Reassess Compliance Posture

The annual compliance review cycle is the heartbeat of a mature AI compliance program. Each year, the compliance function conducts a comprehensive review of the organization's AI systems, compliance documentation, risk classifications, and governance structures against the current state of the regulatory framework. The annual review produces a compliance posture report that identifies gaps between current practice and current requirements, prioritizes remediation actions, and updates the forward-looking compliance roadmap. This report is presented to the AI governance committee and, in summary form, to the board.

Event-triggered reviews supplement the annual cycle. Any significant regulatory development (adoption of a delegated act, publication of final harmonized standards, AI Office enforcement action against a peer organization, material change in the regulatory landscape of a key jurisdiction) should trigger a focused review of the affected compliance areas. Any significant organizational change (acquisition of a company with an AI portfolio, entry into a new market, deployment of a new AI system class, or material change to an existing system) should trigger a risk classification and documentation review for the affected systems. Any significant incident (a serious incident under the Act's definition, a high-profile AI failure in the public domain, or a customer or user complaint that suggests systemic behavior concerns) should trigger a review of the compliance posture for the affected system and any comparable systems in the portfolio.

Governance review cadence should be formally defined in the compliance function's operating model and approved by the AI governance committee. The definition should specify the timing and scope of the annual review, the categories of events that trigger a focused review, and the person responsible for initiating each type of review. A compliance program that conducts reviews on an ad hoc basis, when time permits or when a crisis demands attention, is not operating a governance review cycle; it is operating on a reactive basis that is unlikely to satisfy regulatory expectations for ongoing compliance management.

Diagram 9.9 - Annual Compliance Review Cycle

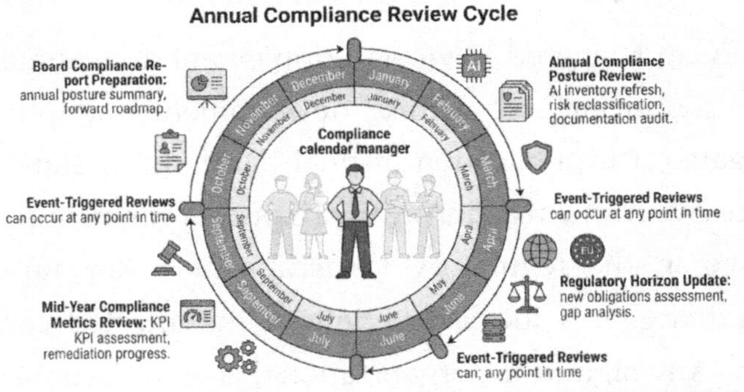

Achieving year-round regulatory alignment through proactive cyclical review.

11.5 Interoperability Across Jurisdictions

For organizations that operate in multiple jurisdictions, the compliance challenge is not just understanding each jurisdiction's requirements in isolation, but building a compliance architecture that satisfies all applicable requirements without creating a separate compliance program for each jurisdiction. The goal of jurisdictional interoperability is a single governance framework that meets the most demanding applicable requirements, supplemented by targeted additions for jurisdictions where specific requirements exceed the global baseline.

Identifying the Compliance Architecture That Satisfies the Most Demanding Applicable Law

The starting point for a globally interoperable compliance architecture is a comparative analysis of applicable AI regulatory requirements across all jurisdictions where the organization operates. This analysis should identify, for each major compliance domain (risk classification, conformity assessment, data governance, human oversight, transparency, incident reporting, post-market monitoring), which jurisdiction imposes the most demanding requirements, and build the compliance architecture to that level. For organizations with significant EU market exposure, the EU AI Act will typically set the high-water mark across most compliance domains, making it the natural foundation for a global compliance architecture.

Comparative analysis reveals both alignment and divergence. In areas of alignment, a single compliance process can satisfy multiple jurisdictions' requirements. In areas of divergence, the compliance architecture must include jurisdiction-specific provisions that address requirements that exceed the global baseline. For example, the EU AI Act's requirement for a CE marking and EU declaration of conformity before market placement of high-risk AI systems is specific to the EU. It has no direct equivalent in most other jurisdictions. However, the underlying activities that support CE marking, including risk assessment, technical documentation, and testing, are broadly aligned with what good AI governance practice requires in other regulatory environments. Building the compliance process to the CE marking standard and then adding a jurisdiction-specific documentation step for the EU formal procedure is more efficient than maintaining parallel processes.

The analysis should also assess where regulatory requirements conflict in ways that create genuine compliance dilemmas. A requirement to provide extensive explainability information to affected individuals under the EU Act may conflict with proprietary information protection claims that a provider might assert under a different jurisdiction's framework. A requirement to retain detailed model performance data for post-market monitoring may conflict with data minimization requirements under privacy laws in specific jurisdictions. These conflicts require legal analysis and, in some cases, jurisdiction-specific compliance positions that the organization is prepared to defend if either jurisdiction's regulator inquires.

Diagram 9.10 - Global Compliance Architecture: Baseline Plus Jurisdictional Modules

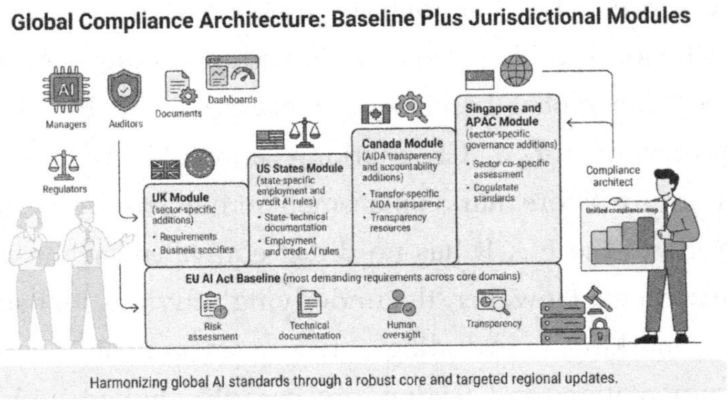

Mutual Recognition, Adequacy Decisions, and Prospects for Regulatory Convergence

Several regulatory frameworks are developing mechanisms to recognize compliance certifications issued in other jurisdictions, thereby reducing the need for duplicative compliance activities. The EU has established adequacy decision-making processes under the GDPR that recognize certain non-EU jurisdictions' data protection frameworks as providing equivalent protection, thereby reducing compliance burdens for data transfers. Comparable mechanisms are being explored for AI governance, and organizations should monitor these developments closely because a mutual recognition arrangement between the EU and a major trading partner could significantly reduce the compliance overhead of operating across both jurisdictions.

The prospects for substantive regulatory convergence are mixed. At the level of high-level principles, there is broad agreement among major AI governance frameworks on the importance of transparency, human oversight, and risk-based regulation. At the level of specific technical requirements, conformity assessment procedures, and enforcement mechanisms, divergence remains substantial. International bodies, including the OECD, the G7 Hiroshima AI Process, and the Global Partnership on AI, are working to develop interoperability frameworks and shared standards that could provide the foundation for greater convergence over time. Still, the timeline and scope of these efforts are uncertain.

The most operationally important form of convergence for enterprise compliance is the alignment between the EU AI

Act's requirements and the international AI governance standards being developed by ISO and IEC. ISO 42001, the international standard for AI management systems, provides a framework that aligns substantially with the EU Act's requirements for organizational governance. Organizations that achieve ISO 42001 certification will have documentation and processes that support EU AI Act compliance without requiring entirely parallel frameworks. Tracking the development and adoption of international AI governance standards and assessing their relationship to applicable legal requirements is a component of the horizon-scanning capability that directly supports compliance architecture decisions.

Building a Globally Coherent AI Compliance Program Without Jurisdiction Silos

The alternative to a globally coherent compliance architecture is a portfolio of jurisdiction-specific compliance programs that operate independently, maintain separate documentation, use different risk classification standards, and are staffed by compliance teams with limited cross-geographic coordination. This siloed approach is inefficient and expensive, and it produces inconsistent compliance outcomes that create audit and enforcement risk when a regulator examines the organization's global AI governance posture and finds that the standards applied in one jurisdiction are materially lower than those in another.

Building a coherent global program requires a governance structure that includes clear ownership of the global compliance architecture, mechanisms for jurisdiction-specific compliance teams to coordinate on shared standards, a unified AI inventory that spans geographies and business units, documentation templates that satisfy the baseline requirements of all applicable jurisdictions, and a compliance technology stack that is accessible and maintained consistently across the organization's geographic footprint. None of these components is technically difficult to build; all of them require sustained organizational discipline to maintain as the organization grows, acquires new businesses, and enters new markets.

The scenario planning dimension of global AI compliance governance deserves explicit attention. Emerging regulatory developments, including the governance of agentic AI systems that can take autonomous actions in the world, regulatory frameworks for synthetic media and deepfakes, and liability regimes for AI systems that operate with significant autonomy in physical environments, will create new compliance obligations that are difficult to predict in detail but certain to arrive in some form. Organizations that build governance structures with the flexibility to absorb new obligation categories, rather than tightly fitting structures to today's specific requirements, will be better positioned to extend their compliance programs to cover emerging regulatory obligations without having to rebuild from the ground up. The responsible-by-design principle is not just about how AI systems are built: it is about how the governance structures around them are built, from the beginning to the end.

11.6 Manager's Checklist

The following items represent the governance and operational capabilities that a manager responsible for AI compliance in a multi-jurisdictional enterprise should be able to confirm. Each item addresses a specific risk associated with managing compliance across a changing regulatory environment.

- Establish a formal regulatory monitoring process with defined information sources, a responsible person, and a regular reporting cycle that produces a documented regulatory intelligence report reviewed by the AI governance committee.
- Conduct an annual comparative analysis of applicable AI regulatory requirements across all jurisdictions where the organization operates, identifying the compliance baseline and any jurisdiction-specific additions required.
- Maintain a compliance documentation architecture using modular, individually versioned documents with change logs, approval records, and a repository that retains all historical versions in a retrievable format.
- Define and publish the event triggers that initiate out-of-cycle compliance reviews, and confirm that the process for initiating such reviews is known to the compliance team, embedded compliance partners, and AI project team leads.
- Assess the organization's participation in regulatory consultation processes and standards development bodies, and determine whether current engagement is

sufficient to represent the organization's interests in the requirements that will govern its AI systems.

- Confirm that the compliance technology stack includes a unified AI inventory that is current, spans all geographies and business units, and that the inventory tracks regulatory designation status for all foundation model dependencies.
- Prepare a scenario analysis addressing at least three emerging regulatory developments (agentic AI governance, synthetic media regulation, autonomous systems liability) and assess their potential impact on the organization's AI portfolio.
- Review the AI governance committee charter at least annually to confirm that its decision rights, composition, and cadence remain appropriate for the organization's current AI compliance obligations and risk profile.
- Confirm that the organization's compliance posture under the EU AI Act is documented in a form that the AI Office could review in an examination, including evidence of the ongoing monitoring, annual review, and adaptive governance processes described in this chapter.

11.7 Takeaway

The organizations that will navigate the next decade of AI regulation most effectively are those that make peace with the central fact of the regulatory environment: it will not stop changing. The EU AI Act will be amended. The European AI Office will issue guidance that clarifies and, in some cases, extends the requirements of the Act. New jurisdictions will adopt frameworks that impose obligations not covered by the EU Act. Emerging technologies, including autonomous AI agents, synthetic media systems, and AI-enabled physical systems, will create governance questions that today's frameworks do not yet answer. None of this is a reason to delay developing a compliance program. It is a reason to build one that is designed for change from the outset.

The governance structures described in this chapter, horizon-scanning capabilities, adaptive compliance processes, modular documentation, annual review cycles, global compliance architectures, and active regulatory engagement, are not compliance overhead. They are the operational infrastructure that enables an organization to convert regulatory awareness into regulatory resilience. A manager who builds these structures is not just managing today's compliance obligations. They are building an organizational capability that compounds in value as the regulatory environment evolves and as competitors who treated compliance as a project fall further behind. That is what it means to govern through change.

12 Closing

Every leader who has worked through the chapters of this book has, in a sense, completed a journey that most of their peers have not yet begun. The EU AI Act, Colorado's AI Act, Canada's AIDA, and the other laws now entering force represent something genuinely new in the history of enterprise technology governance: binding, enforceable, extraterritorial rules that reach inside the organization and assign responsibility for AI systems to named individuals and governing bodies. That is a different kind of obligation than anything most enterprises have encountered in the technology domain, and recognizing its weight is the first act of serious leadership.

This book began with the shift from voluntary principles to enforceable law, because that shift defines the operating environment every manager now inhabits. The voluntary era gave organizations the freedom to interpret AI ethics on their own terms, move at their own pace, and treat governance as a reputational investment rather than a legal obligation. That era is over. The EU AI Act, with its tiered risk framework, conformity assessment requirements, post-market monitoring obligations, and penalties for noncompliance, has transformed the question of AI governance from one of organizational values to one of organizational capability. The question is no longer whether to govern AI responsibly; it is whether the organization can demonstrate that it does.

The risk classification framework at the heart of the EU AI Act is not a bureaucratic artifact. It is a structured judgment about where AI errors cause the most harm, and it has a direct workflow-level impact on every product team, procurement function, and compliance operation in scope. High-risk classification triggers a set of obligations (quality management systems, data governance documentation, human oversight design, post-market monitoring, and conformity assessment) that require deliberate organizational investment to satisfy. Understanding which of your AI systems fall into which tier and maintaining that classification as systems evolve and are modified are among the most foundational and ongoing tasks the Act assigns to enterprise leaders. The substantial modification rule means that classification is not a one-time exercise; it is a live responsibility.

The global regulatory landscape, covered in depth in Chapter Three, adds complexity that many organizations underestimated when they initially focused solely on the EU AI Act. Colorado's algorithmic impact assessment obligations, Canada's AIDA requirements for high-impact systems, Brazil's sector-specific framework, and China's layered approach to algorithmic and generative AI regulation each impose distinct obligations on organizations operating in those markets. The absence of a single federal AI law in the United States has produced fragmentation rather than simplicity, and that fragmentation will continue to grow as additional states follow Colorado's lead. The organizations that will manage this landscape most effectively are those that build a unified compliance architecture capable of absorbing multiple frameworks, not those that treat each jurisdiction as a separate problem requiring a separate solution.

Compliance, as an organizational discipline, is the theme of Chapter Four and perhaps the most important reframing this book asks managers to make. The natural tendency of organizations under regulatory pressure is to respond with documentation: policies, procedures, and written commitments that satisfy the letter of an audit request without changing how work actually gets done. That approach produces the compliance implementation gap, where paper conformity and operational reality diverge. The EU AI Act's requirements are specifically designed to close that gap. Conformity assessment demands technical evidence, not just policy statements. Post-market monitoring requires active surveillance of deployed systems, not just a plan on a shelf. Human oversight must be meaningful, not pro forma. Meeting these requirements demands that compliance be built into the workflows, development processes, and procurement decisions that constitute daily operations. Responsible by design is not a slogan; it is a structural property that organizations either build or fail to build.

The conformity assessment process, addressed in Chapter Five, is where the compliance obligation becomes most concrete and most demanding for organizations with high-risk AI systems. Reverse-engineering what your organization should have been doing all along is how one regulatory affairs director described the experience of preparing for a first conformity assessment. The self-assessment pathway is available for most high-risk systems, but it imposes serious requirements for documentation, testing records, and quality management processes. Third-party assessment is mandatory for systems in the most sensitive categories, and the market capacity of notified assessment bodies is limited. Organizations that wait until they need assessment to build their documentation and quality management practices will face delays, remediation costs, and potential market access problems. The pilot-to-production pathway for a high-risk AI system must now include conformity assessment as a structured gate, not an afterthought.

High-risk AI systems, examined in detail in Chapter Six, represent the compliance frontier where the stakes are highest and the obligations most specific. The Act's Annex III categories cover precisely the domains where AI failures have caused documented harm: hiring, credit, education, law enforcement, critical infrastructure, and judicial proceedings. For organizations that build, deploy, or procure systems in these categories, compliance obligations are non-negotiable and entail high operational costs. Quality management systems, accuracy and robustness requirements, cybersecurity standards, and post-market monitoring plans all must be in place before a high-risk system goes live. The post-market monitoring obligation is particularly worth emphasizing because it extends compliance beyond the deployment decision and into the system's operational life, requiring ongoing attention to performance, drift, and incident reporting.

General-purpose AI and foundation models, the subject of Chapter Seven, introduce a layer of complexity that affects a growing category of enterprises: those that neither develop foundation models nor use consumer AI tools, but build products and services on top of foundation-model APIs. The Act's treatment of this category is still being interpreted, but its direction is clear. GPAI providers face baseline transparency and documentation obligations. Those designated as posing systemic risk face adversarial testing, incident reporting, and heightened cybersecurity requirements. Downstream deployers face questions of shared liability that require careful contractual discipline and ongoing due diligence. As foundation models become more deeply embedded in enterprise software, the compliance questions they raise will become more urgent, not less.

Building the compliance function, the focus of Chapter Eight, is where organizational design choices become compliance outcomes. The roles an organization creates, the reporting structures it establishes, the coordination mechanisms it puts in place, and the budget it allocates all determine whether compliance is a real operational capability or merely nominal. Centralized and federated models each have tradeoffs, and the right choice depends on the organization's size, AI portfolio, risk profile, and existing governance infrastructure. What is not optional is the functional capability itself: the ability to maintain an accurate AI system inventory, conduct or coordinate conformity assessments, monitor deployed systems, engage with regulators, and advise business leaders on the compliance implications of new AI initiatives. Without those capabilities, the governance structure is a diagram rather than a function.

Chapter Nine's argument about adaptive governance most directly challenges the compliance-as-project mentality that still dominates many organizations' thinking. Achieving compliance at a fixed point in time is not the same as solving the compliance problem. The EU AI Act will be amended as implementation experience accumulates. The European AI Office will issue guidance, codes of practice, and delegated acts that add specificity to the Act's requirements. US state laws will proliferate and evolve. International bodies will continue working on interoperability frameworks. Organizations that treat compliance as a project with a completion date will find themselves perpetually behind. Those that treat it as a continuous operational capability (with horizon-scanning functions, adaptive governance processes, and the agility to update their compliance posture as requirements change) will find that the investment compounds into a genuine strategic asset.

There is also something worth noting that extends beyond the compliance obligation itself. The disciplines that AI regulation demands (rigorous documentation, human oversight design, bias and accuracy assessment, post-market monitoring) are not merely regulatory requirements; they are good engineering and good management practice. Organizations that build these disciplines into their AI development and deployment processes will produce better AI systems: systems that perform more reliably, fail more safely, and maintain the trust of the people who use and are affected by them. The regulatory moment and the quality imperative are pointing in the same direction. Compliance work and mission-aligned AI development are not tense; they are the same project.

Senior leaders who have finished this book should have a clear sense of what they need to do next. The gap assessment comes first: an honest inventory of where the organization's AI systems stand relative to applicable regulatory requirements, where documentation and governance gaps exist, and what the compliance risk exposure looks like in concrete terms.

From that foundation, the organizational build-out follows: roles, processes, technology infrastructure, and budget. And from that foundation, the horizon-scanning capability takes shape, so that the organization is always looking ahead rather than catching up.

The enterprises that will lead in the AI economy of the next decade are not necessarily those with the most sophisticated AI systems. They are those with the governance to deploy AI responsibly, the compliance infrastructure to operate across global regulatory environments, and the organizational culture that treats accountability for AI as a mark of institutional quality rather than a constraint on innovation. That is the governed enterprise, and building it requires exactly the kind of mission-aligned leadership this book was written to support. The regulatory moment will not wait. The capability to meet it is built one decision, one process, and one governance structure at a time. That work starts now, and it is yours to lead.

13 References

Access Now. EU AI Act: Civil Society Analysis of High-Risk Obligations. Access Now, 2024. (CH2)

AlgorithmWatch. Regulatory Implications of High-Risk AI Systems in the EU. AlgorithmWatch, 2023. (CH6)

Artificial Intelligence Advisory Council of Canada. *AIDA Implementation Guidance for Enterprises.* Government of Canada, 2024. (CH3)

Baker McKenzie. *Global AI Regulatory Tracker 2025.* Baker McKenzie, 2025. (CH3)

Binns, Reuben, et al. "Accountability in Automated Decision-Making Under EU Law." *International Data Privacy Law,* vol. 13, no. 1, 2024. (CH1)

Brussels European Commission. *Harmonized Standards for the EU AI Act.* European Commission, 2024. (CH5)

Calo, Ryan, and Margot Kaminski. "AI Governance After the EU AI Act." *Yale Journal on Regulation,* vol. 41, 2024. (CH1)

Center for Data Innovation. Comparing Global AI Regulations: EU, US, China, Brazil. CDI, 2023. (CH3)

Centre for European Policy Studies. *Conformity Assessment Under the EU AI Act.* CEPS, 2024. (CH5)

Clifford Chance. EU AI Act: Obligations for High-Risk Providers and Deployers. Clifford Chance, 2024. (CH6)

Council of Europe. Framework Convention on AI, Human Rights, and Democracy. Council of Europe, 2024. (CH9)

Data Protection Authority of France (CNIL). *AI Risk Classification and Compliance Guidance.* CNIL, 2023. (CH2)

Deloitte. Operationalizing AI Compliance Across Global Jurisdictions. Deloitte, 2024. (CH4)

European Commission. *EU AI Act: Final Compromise Text.* Publications Office of the EU, 2024. (CH2)

European Data Protection Board. *Interplay Between GDPR and the EU AI Act.* EDPB, 2024. (CH1)

European Parliament Research Service. *AI Act Enforcement and Supervisory Architecture.* EPRS, 2024. (CH1)

Future of Privacy Forum. General-Purpose AI and Systemic Risk Obligations. FPF, 2024. (CH7)

Gartner. AI Compliance Maturity Models for Enterprise Leaders. Gartner, 2025. (CH8)

Harvard Kennedy School. AI Regulation in the United States: Fragmentation and Federalism. HKS, 2023. (CH3)

Hogan Lovells. EU AI Act: Conformity Assessment and CE Marking Explained. Hogan Lovells, 2024. (CH5)

International Association of Privacy Professionals. *Global AI Governance Report 2024.* IAPP, 2024. (CH9)

International Organization for Standardization. *ISO/IEC 42001: AI Management System Standard.* ISO, 2023. (CH4)

KPMG. Building AI Compliance Functions in Large Enterprises. KPMG, 2024. (CH8)

McKinsey & Company. AI Regulation and Enterprise Readiness 2025. McKinsey, 2025. (CH4)

National Institute of Standards and Technology. *AI Risk Management Framework (AI RMF 1.0).* NIST, 2023. (CH4)

OECD. OECD Framework for Classification of AI Systems. OECD, 2023. (CH2)

OpenAI. GPAI Transparency and Safety Disclosures. OpenAI, 2024. (CH7)

PwC. EU AI Act Readiness Assessment for Multinationals. PwC, 2024. (CH4)

Stanford Institute for Human-Centered AI. *Global AI Index 2024.* Stanford HAI, 2024. (CH3)

UK Information Commissioner's Office. *AI and Data Protection Risk Mitigation Guidance.* ICO, 2023. (CH6)

United Nations. *Global AI Governance Framework 2024.* United Nations, 2024. (CH9)

World Economic Forum. AI Governance and Cross-Border Compliance. WEF, 2024. (CH9)

World Economic Forum. *Responsible AI for High-Risk Sectors.* WEF, 2023. (CH6)

Zhang, Lin, et al. "Regulatory Approaches to Foundation Models." *AI and Ethics,* vol. 5, 2024. (CH7)

Made in United States
Orlando, FL
04 November 2026

90201056R00164